Introduction to Nonlinear Laser Spectroscopy

QUANTUM ELECTRONICS—PRINCIPLES AND APPLICATIONS

A Series of Monographs
EDITED BY

PAUL F. LIAO
Bell Telephone Laboratories
Murray Hill, New Jersey

PAUL KELLEY
Lincoln Laboratory
Massachusetts Institute of Technology
Lexington, Massachusetts

N. S. Kapany and J. J. Burke. OPTICAL WAVEGUIDES, 1972

Dietrich Marcuse. THEORY OF DIELECTRIC OPTICAL WAVEGUIDES, 1974

Benjamin Chu. LASER LIGHT SCATTERING, 1974

Bruno Crosignani, Paolo Di Porto, and Mario Bertolotti. STATISTICAL PROPERTIES OF SCATTERED LIGHT, 1975

John D. Anderson, Jr. GASDYNAMIC LASERS: AN INTRODUCTION, 1976

W. W. Duley. CO_2 LASERS: EFFECTS AND APPLICATIONS, 1976

Henry Kressel and J. K. Butler. SEMICONDUCTOR LASERS AND HETEROJUNCTION LEDs, 1977

H. C. Casey and M. B. Panish. HETEROSTRUCTURE LASERS: PART A, FUNDAMENTAL PRINCIPLES: PART B, MATERIALS AND OPERATING CHARACTERISTICS, 1978

Robert K. Erf (Ed.). SPECKLE METROLOGY, 1979

Marc D. Levenson. INTRODUCTION TO NONLINEAR LASER SPECTROSCOPY, 1982

In Preparation

Robert A. Fisher (Ed.). OPTICAL PHASE CONJUGATION, 1982

INTRODUCTION TO NONLINEAR LASER SPECTROSCOPY

Marc D. Levenson
IBM Research Laboratory
San Jose, California

1982

ACADEMIC PRESS
A Subsidiary of Harcourt Brace Jovanovich, Publishers

New York   London
Paris   San Diego   San Francisco   São Paulo   Sydney   Tokyo   Toronto

COPYRIGHT © 1982, BY ACADEMIC PRESS, INC.
ALL RIGHTS RESERVED.
NO PART OF THIS PUBLICATION MAY BE REPRODUCED OR
TRANSMITTED IN ANY FORM OR BY ANY MEANS, ELECTRONIC
OR MECHANICAL, INCLUDING PHOTOCOPY, RECORDING, OR ANY
INFORMATION STORAGE AND RETRIEVAL SYSTEM, WITHOUT
PERMISSION IN WRITING FROM THE PUBLISHER.

ACADEMIC PRESS, INC.
111 Fifth Avenue, New York, New York 10003

United Kingdom Edition published by
ACADEMIC PRESS, INC. (LONDON) LTD.
24/28 Oval Road, London NW1 7DX

Library of Congress Cataloging in Publication Data

Levenson, Marc D.
  Introduction to nonlinear laser spectroscopy.

  (Quantum electronics—principles and applications)
  Includes bibliographical references and index.
  1. Laser spectroscopy. 2. Nonlinear optics. I. Title.
II. Series.
QC454.L3L48   535.5'8   81-17608
ISBN 0-12-444720-1   AACR2

PRINTED IN THE UNITED STATES OF AMERICA

82 83 84 85   9 8 7 6 5 4 3 2 1

For my father

CONTENTS

Foreword   xi
Preface   xiii

1. Introduction

    1.1 Prologue: Linear Spectroscopy   1
    1.2 The Tunable Laser   5
    1.3 A Catalog of Nonlinear Phenomena   13
    1.4 Linear Spectroscopy with Nonlinear Sources   25
    1.5 Laser Spectroscopy Literature   26
           References   27

2. Theory

    2.1 The Density Matrix for a Two-Level System   29
    2.2 The Interactions and the Hamiltonian   32
    2.3 Relaxation   32
    2.4 The Master Equation and the Vector Model   34
    2.5 The Nonlinear Polarization Density and Nonlinear Susceptibility   45
    2.6 Inhomogeneous Broadening   47
    2.7 Effective Operators for Multiquantum Transitions   50
    2.8 Multiple Resonance Effects   54
    2.9 The Wave Equation and the Detected Signal   60

2.10 The Recipe   63
References   64

3. Saturation Spectroscopy

3.1 Burning and Detecting Holes in a Doppler-Broadened Two-Level System   66
3.2 Crossover Resonances and Polarization Spectroscopy   73
3.3 Coupled Doppler-Broadened Transitions   80
3.4 Experimental Methods of Saturation Spectroscopy in Gases   84
3.5 Ramsey Fringes in Saturation Spectroscopy   92
3.6 The Line-Shape Problem in Saturation Spectroscopy   96
3.7 Experimental Results in Saturation Spectroscopy of Gases   97
3.8 Multiphoton and Double-Resonance Saturation Techniques   101
3.9 Saturation Techniques for Condensed Phases   104
3.10 Applications of Saturation Techniques   110
References   112

4. Coherent Raman Spectroscopy

Introduction   115
4.1 Driving and Detecting a Raman Mode   116
4.2 Symmetry Considerations   121
4.3 Relationship between χ^R and the Spontaneous Cross Section   130
4.4 Wave-Vector Matching   130
4.5 Coherent Anti-Stokes Raman Spectroscopy   132
4.6 Raman-Induced Kerr Effect Spectroscopy   139
4.7 Stimulated Raman Gain and Loss Spectroscopy   145
4.8 Four-Wave Mixing   147
4.9 Applications   148
4.10 Judging the Merits: The Signal-to-Noise Ratio   156
References   159

5. Multiphoton Absorption

5.1 Introduction   161
5.2 Doppler-Free Two- and Three-Photon Absorption   164
5.3 Multiquantum Ionization   172
5.4 Nonlinear Mixing   185
5.5 Applications   190
References   192

6. Optical Coherent Transients

    6.1   The Optical Free-Induction Decay   195
    6.2   Optical Nutation   199
    6.3   The Photon Echo   201
    6.4   The Stimulated Echo   206
    6.5   Ramsey Fringes   211
    6.6   Experimental Techniques and Results   217
           References   224

7. Nonlinear Sources for Linear and Nonlinear Spectroscopy

    7.1   Second Harmonic and Sum Frequency Generation   226
    7.2   Third- and Higher-Order Sum and Harmonic Generation   230
    7.3   Raman Shifting   232
    7.4   Spontaneous XUV Anti-Stokes   236
    7.5   Infrared Spectrophotography   237
           References   240

Appendix: Symbol Glossary-Index   241

Index   251

FOREWORD

This book appears at an opportune time, when the field of nonlinear laser spectroscopy has been recognized by the 1981 Nobel prize awards in physics. I had the good fortune to receive a copy of the author's manuscript just in time to assist me in preparing my 1981 Nobel lecture.

The book gives an authoritative, up-to-date account of the principles of nonlinear laser spectroscopy. Many different facets are treated from a unified point of view. A particularly valuable feature, which is not common in books of this kind, is the thorough discussion and comparison of various experimental approaches. The abundant references provide ready access to the original research literature.

As the field of nonlinear laser spectroscopy has sufficiently matured during the past decade, the danger of rapid obsolescence for this book is small. The subject matter, however, still offers plenty of opportunity for further exploration and exploitation. Many new workers in other branches of science and technology may be attracted to it, and they will need the introduction to the field that this volume provides. In my opinion this book will make a substantial contribution to the further growth of an increasingly important subfield of physics.

N. BLOEMBERGEN

The past twenty years, since the advent of lasers, have seen the beginnings and enormous growth in the science of nonlinear optics. No longer do light beams move through materials simply being absorbed, reflected, or refracted. A wide range of new phenomena, including the production of new wavelengths by harmonic generation and mixing, occurs at high light intensities.

As lasers have become more widely tunable, the field has blossomed into nonlinear spectroscopy. Very small traces of materials can be seen by nonlinear techniques such as Coherent Anti-Stokes Raman Spectroscopy (CARS). Sharp-line spectra free from the Doppler broadening of thermal motions can now be obtained in a wide variety of ways, again using the nonlinear effects from intense laser light.

Dr. Levenson has been one of the leading pioneers in this field. At Stanford University, Massachusetts Institute of Technology, the University of Southern California, and now at the IBM Research Laboratory, he has played a major role in introducing many of the most important new experimental techniques of nonlinear optics and spectroscopy. He is uniquely well qualified to present them to the scientific community in this book.

A. L. SCHAWLOW

PREFACE

The development of powerful and convenient tunable lasers has revolutionized optical spectroscopy. A wide variety of techniques have been developed in the past 20 years that exploit the ability of these high spectral brightness sources to manipulate population distributions and induce multiquantum transitions. These nonlinear spectroscopic processes are qualitatively different from the processes employed in conventional optical spectroscopy and are more closely related to nuclear magnetic resonance and electron spin resonance. The apparent diversity of the nonlinear techniques masks an underlying unity.

My purpose in writing this book is to unify the presentation of the most useful nonlinear spectroscopy techniques at a level accessible to graduate students and spectroscopists unfamiliar with nonlinear optics. Some familiarity with laser physics and an elementary understanding of quantum mechanics are assumed. The dynamics of the nonlinear optical process is emphasized at the expense of the physics of the spectra being measured. While the optical nonlinear susceptibility for a general quantum system is introduced, most of the calculations are in terms of an effective two-level model. The very recent developments and phenomena that intrinsically require more than two resonant levels have been omitted in the interest of clarity.

Many people have contributed to the publication of this volume. I would like especially to acknowledge my indebtedness to the many authors whose work I have quoted, some of whom supplied me with historically important figures. I acknowledge the support of the IBM Corporation and my many colleagues at IBM, especially J. D. Swalen, E. M. Engler, and Dietrich Haarer. I extend special thanks to Tracy Takagi and Karen Bryan for their assistance in preparing the typescript and figures, and to Jean Chen and Frank Schellenberg for their assistance in designing the cover. I also acknowledge the inspiration provided by mentors and former colleagues, especially A. L. Schawlow, T. W. Hansch, N. Bloembergen, R.W. Hellwarth, and R. G. Brewer.

Chapter 1
INTRODUCTION

1.1 PROLOGUE: LINEAR SPECTROSCOPY

In 1814, Josef Fraunhofer used new and more precise apparatus to repeat an experiment performed by Newton more than a century before: he dispersed the solar spectrum into its component colors. Fraunhofer's spectroscope had greater resolution than those employed previously, and amid the familiar wash of color, he found a new effect: narrow dark lines appeared in the solar spectrum with a definite and unchanging pattern. Spectra of other light sources—notably flames—showed similar structures, often with bright lines. When table salt was shaken into a dark flame, the bright orange light had the exact same wavelength as two prominent dark lines in the solar spectrum.

Thus began spectroscopy as we now know it. Within 50 years, certain dark and bright lines were known to be characteristic of the chemical elements. It was assumed that the energy in light could somehow excite the internal vibrations of an atom, leading to absorption. Similarly, heat or electricity could excite vibrations which would radiate the energy away as light. Such classical models as refined by Lorentz accounted for the coupling between light and matter [1].

When Maxwell formulated the electromagnetic theory of light, he made the approximation that the dielectric susceptibility and magnetic permeability were independent of the strengths of the applied fields, and thus the dielectric polarization and magnetization were linearly proportional to the field amplitudes,

$$\mathbf{P} = \vec{\chi} \cdot \mathbf{E}; \qquad \mathbf{M} = (\vec{\mathscr{K}} - \vec{1}) \cdot \mathbf{H}. \tag{1.1.1}$$

The resonant structure of the spectra were contained in the susceptibilities. More modern workers allowed the susceptibilities to be regarded as complex numbers and the optical frequency fields to be written in complex notation

$$\mathbf{E}(\mathbf{r}, t) = \operatorname{Re} \mathbf{E}(\mathbf{r})e^{-i\omega t} = \tfrac{1}{2}\{\mathbf{E}(\mathbf{r})e^{-i\omega t} + \mathbf{E}^*(\mathbf{r})e^{i\omega t}\}, \tag{1.1.2}$$

with the operator "Re" generally omitted. The famous wave equation thus became

$$\nabla^2 \mathbf{E} - \frac{1}{c^2}\frac{\partial^2}{\partial t^2}\mathbf{E} = \frac{4\pi}{c^2}\frac{\partial^2}{\partial t^2}\mathbf{P} \tag{1.1.3}$$

which predicted a phase velocity in nonmagnetic media of

$$v = cn^{-1} = c\operatorname{Re}(1 + 4\pi\chi)^{-1/2}, \tag{1.1.4}$$

where n is the usual index of refraction. The attenuation in absorbing media obeyed Beer–Lambert's law,

$$(\hat{k} \cdot \nabla)I = -\kappa I, \tag{1.1.5}$$

where the unit vector $\hat{k}$ specifies the local direction of propagation, the attenuation constant

$$\kappa = 2\omega c^{-1} \operatorname{Im}(1 + 4\pi\chi)^{1/2}, \tag{1.1.6}$$

and the observable intensity of the wave (in ergs cm^{-2} sec^{-1}) was related to the time averaged Pointing vector as

$$I = \hat{k} \cdot \langle \mathbf{S} \rangle = (nc/4\pi)|\mathbf{E}(\mathbf{r})|^2. \tag{1.1.7}$$

The early workers in quantum mechanics recognized that variation of the index of refraction and attenuation coefficient contained crucial information as to the energy levels of the medium. Bohr related the attenuation coefficient to the probability of a transition between energy levels separated by the energy of quantum of light $E = \hbar\omega$. Fermi expressed the transition rate in terms of the matrix element of the dipole moment operator $\boldsymbol{\mu} = e\mathbf{r}$ which connected the two levels

$$\Gamma_{ij} = (2\pi/\hbar)|\langle i|\boldsymbol{\mu} \cdot \mathbf{E}(\mathbf{r})|j\rangle|^2 \rho(E_i - E_j - \hbar\omega) = (n/c)B_{ij}I(\omega), \tag{1.1.8}$$

where the factor $\rho(E)$ is a density of states function that reflects the observed lineshape and e is the electronic charge. Actually, when the index of refraction of the medium is different from one, the dipole moment operator must be corrected for the local field and $\boldsymbol{\mu} \to ((n^2 + 2)/3)e\mathbf{r}$. Einstein related the

1.1 Prologue: Linear Spectroscopy

probability of emission to that for absorption or "stimulated emission"

$$A_{ij} = (2\hbar\omega^3/\pi c^3)B_{ij}, \tag{1.1.9}$$

and Kramers and Kronig showed that the variation of the index of refraction was related to the absorption

$$\text{Re}\,\chi(\omega) = \frac{1}{\pi}\int_{-\infty}^{\infty}\frac{\text{Im}\,\chi(\omega')}{\omega'-\omega}d\omega';$$

$$\text{Im}\,\chi(\omega) = -\frac{1}{\pi}\int_{-\infty}^{\infty}\frac{\text{Re}\,\chi(\omega')}{\omega'-\omega}d\omega'. \tag{1.1.10}$$

Thus, the quantum mechanics of the classical tools of spectroscopy—emission, absorption, and dispersion—was well understood early in the twentieth century [2–4, 21].

The fourth tool of linear spectroscopy—light scattering—required a bit more sophistication. It was evident, however, that the models of a molecule in which the electronic forces holding the atoms together were treated as springs would permit the nuclei to vibrate at frequency Ω_Q and that this vibration might modulate the dielectric constant. If Q is the generalized coordinate that describes the nuclear vibration, the polarizability of a single vibrating molecule can be expressed as

$$\alpha(t) = \alpha_0 + \frac{\partial\alpha}{\partial Q}Q(t), \tag{1.1.11}$$

and the dielectric polarization due to this molecule illuminated by a light wave of amplitude **E** and frequency ω is

$$\mathbf{P}(t) = \alpha(t)\mathbf{E}(t) = \alpha_0\mathbf{E}(\mathbf{r})e^{-i\omega t}$$
$$+ \frac{\partial\alpha}{\alpha Q}\mathbf{E}(\mathbf{r})Q_0\{e^{-i(\omega+\Omega_Q)t} + e^{-i(\omega-\Omega_Q)t}\}. \tag{1.1.12}$$

The terms in Eq. (1.1.12) oscillating at $\omega \pm \Omega_Q$ radiate waves at these frequencies according to Eq. (1.1.3). The waves scattered by different molecules add incoherently, and thus the total scattered power per unit volume is proportional to the incident intensity and the number of molecules per unit volume [5],

$$\mathscr{P}_{\text{scatt}} = \mathscr{N}\sigma_Q I. \tag{1.1.13}$$

The correct expression for the total Raman scattering cross section σ_Q is [6]

$$\sigma_Q = \frac{4\pi\hbar}{9c^4\Omega_Q}\frac{(\omega-\Omega_Q)^4 hg_Q}{[1-\exp(-\hbar\Omega_Q/kT)]}\sum_{\gamma\delta}\left|\left(\frac{\partial\alpha_{\gamma\delta}}{\partial Q}\right)\right|^2. \tag{1.1.14}$$

In Eq. (1.1.14), g_Q is the degeneracy of the Raman mode, and the squares of all the elements or the polarizability tensor derivative are summed [5].

The relationship between the Raman transition polarizability of Chapter 4 and the polarizability derivative used here is

$$\partial \alpha_{\gamma\delta}/\partial Q = (2\Omega_Q/\hbar)^{1/2} \alpha_{\gamma\delta}^R.$$

These classical phenomena of linear spectroscopy contributed immeasurably to the growth of quantum mechanics and to our understanding of atoms, molecules, and crystals. The experimental limitations that resulted from unfavorable selection rules, Doppler broadening, and other complicating phenomena were recognized as intrinsic to linear spectroscopy, and largely tolerated. Before the development of the laser, the interactions between optical frequency fields and matter were weak enough that essentially linear theories sufficed. There was, thus, no strong motivation to develop a more general treatment. That was not the case in the radio-frequency and microwave regions where strong coherent sources became available in the 1930s. Nonlinear effects in dielectric and magnetic media were recognized early and soon exploited technologically [7, 8].

Detailed understanding of nonlinear resonance phenomena began with the discovery and explanation of nuclear magnetic resonance in the late 1940s. The dynamics of most of the interesting nonlinear optical resonances is analogous to that of an ensemble of spin $\frac{1}{2}$ systems. A recent book by Allen and Eberly recounts the effects observed in two-level systems [9].

More generally, the phenomena observed in the steady state can be described by a dielectric polarization density expanded as a power series in the electric field,

$$P_i = \chi_{ij} E_j + \chi_{ijk}^{(2)} E_j E_k + \chi_{ijkl}^{(3)} E_j E_k E_l + \cdots, \qquad (1.1.15)$$

where the subscripts denote Cartesian coordinates. The complex tensor coefficients of the higher-order terms are called nonlinear susceptibilities [10]. The nonlinear optical effects useful in spectroscopy—saturated absorption, stimulated Raman gain, four-wave mixing, multiquantum absorption, sum frequency generation, etc.—result from these terms. Techniques based on these effects widen the range of optical spectroscopy and increase its precision. Doppler broadening, for example, can be eliminated; weak and "forbidden" processes can be enhanced tremendously. New information on the energy levels, lifetimes, collisional processes, coupling strengths, etc. is encoded in the resonant behavior of the nonlinear susceptibilities. These nonlinear resonances and the means of observing them will be the subject of the rest of this monograph. As in the time of Fraunhofer, new technology continues to uncover unexpected phenomena in areas that were thought to be well understood.

1.2 THE TUNABLE LASER

The fundamental tool of laser spectroscopy is a source of coherent light of variable frequency. Today, there is a wide variety of such tunable sources: continuous wave (cw) and pulsed dye lasers, F-center lasers, parametric oscillators, spin-flip Raman lasers, tunable diode lasers, etc., but many of the fundamental laser spectroscopy experiments were done without such convenient apparatus. Actually, any laser can be made tunable by constructing a resonator that allows oscillation over a band narrower than the gain band of the medium. The problem is that the narrow tuning ranges thus obtained do not overlap many interesting material resonances. The laser medium itself can, however, be studied, and thus many of the first nonlinear spectroscopy projects were done on laser media.

Another early tuning strategy was to shift the output of a strong fixed frequency laser with Raman laser oscillators employing various media. By judicious selection of Raman frequency, the output could be adjusted in steps of a few inverse centimeters. Other experiments were done employing combinations of laser and incoherent illumination, or varying angles rather than wavelengths [10].

1.2.1 The Dye Laser

As impressive as these early efforts might be, laser spectroscopy did not become practical until the development of the reliable dye laser. In the dye laser, an intense light source pumps molecules of some organic dye into high vibrational sublevels of the first excited singlet state. From there, a fast relaxation process takes them to the lowest few sublevels, the populations of which then become greater than those of the excited vibrational levels of the ground electronic state. If no other effects siphon away the population, and if a suitable resonator is provided, laser oscillation can occur over a continuous broad range of frequencies as shown in Fig. 1.1. The science and technology of the dye laser has been reviewed in an excellent monograph and several long articles [11]. There is no need to recount that information here, but rather, we shall introduce the topics directly relevant to nonlinear spectroscopy.

The general strategy for converting a laser with a broad gain band into a source of tunable coherent radiation is shown in Fig. 1.2. Several frequency selective elements are usually necessary to sufficiently narrow the output. The coarse tuning element may be a prism, grating, interference filter, or birefringent tuning plate that has low loss only over a few inverse centimeters within the several hundred inverse centimeter wide dye gain band. Laser

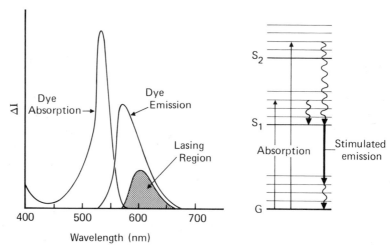

Fig. 1.1 The absorption and emission spectrum of a typical laser dye and the dynamics of stimulated emission. The spectra are of rhodamine 6G. Absorption of pump light populates vibrationally excited levels of the first two excited singlets, which relax rapidly to the lowest excited level. Since the population of this level exceeds that of the vibrationally excited levels of the ground electronic state, stimulated emission produces gain. The lower laser level again relaxes rapidly toward thermal equilibrium. Because the absorption spectrum overlaps the fluorescence spectrum, the region where gain exceeds loss is shifted somewhat to the red of the peak emission, but gain exists over a relatively wide frequency range.

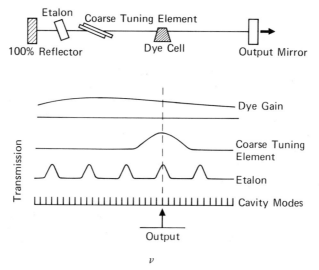

Fig. 1.2 A typical single-mode tunable laser cavity and the principles of operation. The gain available over the wide lasing region of the dye is channeled into a narrow frequency interval by a series of frequency selective elements. The coarse tuning element allows oscillation over roughly $\frac{1}{4}$ Å. The etalon is adjusted to have one passband within this range, narrowing the output further. Within the etalon passband, one mode of the overall cavity will have the minimum loss, and will thus compete most effectively for the available population inversion. Normally, such a laser will oscillate on a single axial mode (or a few nearby modes) with a linewidth of a few megahertz. Tuning is accomplished by altering the etalon, coarse tuning element, and cavity mode frequencies simultaneously.

1.2 The Tunable Laser

oscillation occurs only in this region of low loss, which can be tuned by rotating, tilting, or translating the tuning element.

The output spectrum can be narrowed further with a Fabry–Perot interferometer or etalon having a free spectral range larger than the low-loss band of the coarse tuning element. Oscillation can then take place over a band typically of width 1 GHz or so. The output frequency can be varied by changing the angle of the etalon, the spacing of the plates, or the index of refraction of the medium between the plates. The longitudinal modes of a typical dye laser cavity are spaced by a few hundred megahertz, and thus several modes can oscillate simultaneously within the gain band of this Fabry–Perot interferometer. Occasionally, competition between modes will result in oscillation on a single cavity mode, but if not, a second "fine tuning" etalon can be used to narrow the gain band to less than a cavity mode spacing. The frequency of the cavity mode can be varied by translating one mirror piezoelectrically or by tilting an intracavity "Brewster plate" to vary the optical path length (see Fig. 1.4b).

The output spectrum of a laser oscillating on a single cavity mode can be narrow indeed. The Schawlow–Townes linewidth relation implies a quantum limited linewidth of a millihertz or less for a cw laser well above threshold [12]. More generally, mechanical instabilities induce a frequency jitter of several megahertz. Fast-servo techniques can compensate for these instabilities producing linewidths down to 500 Hz for a cw laser. The narrowest bandwidth achievable in a pulsed laser is set by the Fourier transform of the temporal profile of the pulse. The Fourier transform of a Gaussian pulse with a $1/e$ full width of τ, is a Gaussian line-shape function with a $1/e$ width of $\Delta v = 4/\pi\tau$. Different pulse shapes imply different transform limited linewidths. It is generally quite difficult to achieve the transform limit. If it is not achieved, one must assume that laser amplitude and phase are rapidly and unobservably fluctuating.

The light source pumping the dye laser must be intense enough to compete with the effects that cause the population inversion to decay. While flash lamp pumped dye lasers are suitable for some purposes, the dye laser designs most useful in spectroscopy require pumping by means of another laser. Two designs for pulsed narrow-band dye oscillators appear in Fig. 1.3. In both cases, a diffraction grating acts as one mirror of the laser cavity, reflecting light back into the gain medium only when the wavelength of that light fulfills the Littrow condition

$$\lambda = 2d \sin \theta \tag{1.2.1}$$

where d is the grating spacing and θ the angle between the light beam and the normal to the grating surface.

The resolution of a Littrow mount grating depends upon the width of the illuminated region. Hence, both laser designs incorporate some means

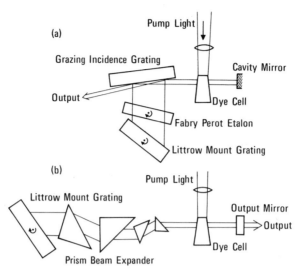

Fig. 1.3 Two cavity designs for narrow-band pulsed lasers. (a) In the Littman design, a grazing incidence grating expands the beam in the horizontal direction and directs it through an etalon into a Littrow mount grating that acts as a coarse tuning element. (b) A Hänsch–Klauminzer design in which the expansion necessary to realize the full resolution of the grating is achieved using prisms.

of expanding the narrow beam emanating from the dye cell into a wider collimated beam at the grating [13]. In one case, another diffraction grating used near grazing incidence performs this task. The frequency of the laser is varied by rotating the gratings as shown. The output spectrum can be narrowed further by inserting a coated plane Fabry–Perot interferometer into the cavity as shown in Fig. 1.3a. Linewidths as small as 15 MHz have been obtained in this manner, but tuning then requires synchronous adjustment of the grating, interferometer, and overall cavity length. Such sychronization can be elegantly accomplished by immersing the entire device in a variable pressure chamber where the gas density (and hence index of refraction) can be scanned.

Two designs for cw dye lasers appear in Fig. 1.4. In each case, the high pumping rate required to achieve population inversion is obtained by focusing several watts of TEM_{00} power from an ion laser to a spot a few microns in diameter within the dye medium. The heat thus deposited must be transported away by flowing the dye medium at a speed of several meters per second. This is now conventionally accomplished with a free-standing jet of high viscosity dye medium. The simple standing wave cavity in Fig. 1.4a is suitable for multimode operation or single mode operation at low power. The folded three-mirror design allows the beam waist of the dye laser cavity

1.2 The Tunable Laser

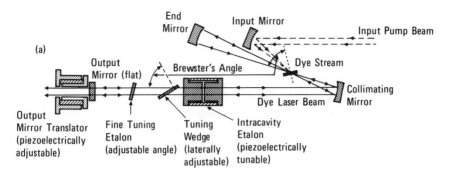

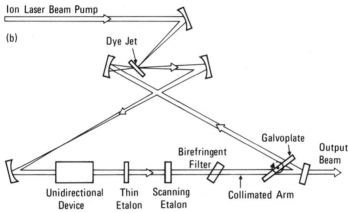

Fig. 1.4 Two designs for single-mode cw dye lasers. (a) The standing wave cavity works best at low power. The tuning wedge acts as a coarse tuning element and the two intracavity etalons narrow the spectrum to a single axial mode. Without these etalons, such a cavity will produce output efficiently over a 1-Å bandwidth. (b) The ring-laser cavity is more efficient for single-mode operation at high power. The unidirectional device employs a Faraday rotator to discourage oscillation in the direction opposite the arrows. The birefringent filter acts as coarse tuning element, and the axial mode frequencies are adjusted by galvanometrically tipping a glass plate oriented near Brewster's angle.

to overlap the pumped region. Coarse tuning is by means of a variable spacing interference filter tilted at Brewster's angle. In single-mode operation, a standing wave cavity such as in Fig. 1.4a will produce nodal planes within the dye medium where the intensity of the dye laser light is zero. Molecules in these planes cannot contribute to the desired output frequency, but can produce gain on nearby modes. At high pump intensities, this spatial hole burning effect leads to disastrous instabilities.

The ring cavity laser design shown in Fig. 1.4b produces high powers in a single frequency, but at the price of increased complexity. As in the previous case, folding mirrors produce a small beam waist where the pump beam strikes the dye stream. The dye laser radiation resonates in a bowtie-shaped cavity. A unidirectional device incorporating a Faraday rotator and (usually) an optically active quartz plate ensures that the light propagating in one direction encounters lower losses than that propagating in the opposite direction. No nodes due to standing wave effects can appear, and the laser thus operates stably in a single mode. Coarse tuning is accomplished with a stack of birefringent plates tilted at Brewster's angle. Only certain wavelengths propagate through such a stack without polarization change sufficient to cause loss. When the stack is rotated about an axis normal to the surface, these low loss wavelengths vary. An etalon with piezoelectrically variable plate spacing selects the desired cavity mode. A Brewster-angled tuning plate near one folding mirror varies the mode frequency without excessively altering the output beam direction. Servo control and synchronization is necessary to tune smoothly.

The organic dyes most suitable as laser media in various wavelength regions are listed in Table 1.1. The list is not meant to be complete, but does indicate that these sources can cover the range from roughly 3500 Å to

TABLE 1.1 *The Most Useful Laser Dyes*

Laser Dye[a]	Wavelength (nm)	Solvent[b]	Pump[c]
p-Terphenyl	323-364	C_6H_{12}	KrF
Butyl-PBD	355-390	DMF	Lamp
PBD	360-386	ETOH/Toluene	N_2
BBQ	370-410	DMF	Lamp
α-NPO	385-415	C_6H_{12}	N_2
Stilbene 3	401-490	ETG	Ar^+(UV-CW)
7-Diethylamino -4-Methyl Coumarin	446-506	ETG	Ar^+(UV-CW)
Coumarin 30	495-545	ETG	Ar^+(CW)
Sodium Fluorescein	538-573	ETG	Ar^+(CW)
Rhodamine 110	540-600	ETG	Ar^+(CW)
Rhodamine 6G	570-650	ETG	Ar^+(CW)
DCM	610-730	ETG/DMSO	Ar^+(CW)
Oxazine 1	690-780	ETG	Kr^+(CW)
LD 700	720-860	ETG	Kr^+(CW)
HITC	835-955	DMSO	Kr^+(CW)
IR-140	890-980	DMSO	Kr^+(IR-CW)

[a] Available from Eastman Kodak Corporation, Exciton Chemical Company, or Lambda Physik GmbH.

[b] Solvent abbreviations: C_6H_{12}, cyclohexane; DMF, dimethyl formamide; ETOH, ethyl alcohol; ETG, ethylene glycol; DMSO, dimethyl sulfoxide.

[c] Only the weakest workable pump is indicated.

1.2 The Tunable Laser

roughly 9000 Å. Not all of these molecules are stable, and few dyes approach rhodamine 6G in convenience. Harmonic generation crystals are capable of doubling or summing these wavelengths to produce useful wavelengths down to roughly 2500 Å. Alternatively, the output from a master oscillator can be amplified in a pulsed traveling-wave amplifier chain or a regenerative injection-locked amplifier to obtain megawatt power levels [14, 15].

1.2.2 Other Tunable Sources

Additional tunable laser systems have been developed to probe spectroscopic phenomena that occur outside the wavelength range spanned by the organic dyes. Most similar to the dye media are the color center lasers demonstrated in the alkali halides [16]. The presently available wavelengths are listed in Table 1.2. In these media, electrons trapped at lattice vacancies are excited by the pump radiation, "relax" by distorting the vacancy sites, and thus produce a population inversion with respect to the "relaxed" ground state. Laser action occurs over a wide gain band shifted to longer wavelength from the absorption band. Cavity designs similar to Figs. 1.3 and 1.4 can be employed with an alkali halide crystal at liquid nitrogen

TABLE 1.2 *Color Centers Useful in Tunable Lasers*[a]

Host	Center	Wavelength Range (nm)	Pump[c] (CW)
LiF	F_2^+	820-1050	Kr^+
NaF	$(F_2^+)^*$	990-1220	$LiF:F_2^+$
KF	F_2^+	1220-1500	Nd:YAG
NaCl	F_2^+	1400-1750	Nd:YAG
KCl:Na	$(F_2^+)_A$	1620-1910	Nd:YAG (1.34 μm)
KCl:Li	$(F_2^+)_A$	2000-2500	Nd:YAG (1.34 μm)
KCl:Na[b]	$F_B(II)$	2250-2650	Kr^+
KCl:Li[b]	$F_A(II)$	2500-2900	Kr^+
RbCl:Li[b]	$F_A(II)$	2600-3300	Kr^+

[a] Data from Linn F. Mollenauer in CRC Laser Handbook.
[b] Available from Burleigh Corporation.
[c] Only the weakest useable pump is indicated.

temperature substituted for the dye stream. Suitable crystals can be prepared and "refreshed" by various chemical and photochemical processes.

Another well-developed tunable source is the parametric oscillator which makes use of the nonlinear optical process to split pump photons into a "signal" and "idler" photon which together conserve energy and wave vector [17]. The band over which this process is effective depends upon the phase-matching condition

$$\mathbf{k}_p = \mathbf{k}_s + \mathbf{k}_I, \qquad (1.2.2)$$

where $|\mathbf{k}_i| = n_i \omega_i / c$, and the subscripts refer to the pump, signal, and idler beams, respectively.

Tuning is possible by varying the angle and temperature of the parametric oscillator crystal. In Fig. 1.5 are shown the signal and idler wavelengths generated by a LiNbO$_3$ parametric oscillator pumped by a 1.06 μm Nd:YAG laser as a function of the angle between the optical axis of the crystal and the propagation direction. The output bandwidth must, however, be narrower than the gain bandwidth for this device to be useful in spectroscopy. Thus, cavity structures similar to Fig. 1.3 are employed, but modified to allow collinear propagation of pump and output beams.

Further in the infrared, the lead salt diode lasers produce narrow output lines at variable frequencies, but with too little power to be useful in most

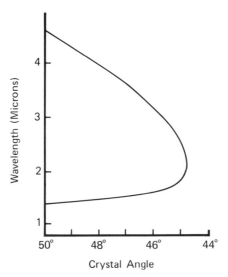

Fig. 1.5 Wavelengths produced by a LiNbO$_3$ parametric oscillator pumped by 1.06-μm radiation; $T = 25°$C. The crystal angle is the angle between the beam direction in the parametric oscillator cavity and the optical axis of the crystal. Two wavelengths—the signal and idler—are produced, and the sum of their frequencies equals the pump frequency.

nonlinear spectroscopy processes [18]. Nevertheless, commercially available sources of this type make high-resolution spectroscopy possible over the range 4.8–28 μm. Spin-flip Raman lasers pumped by CO and CO_2 lasers are more powerful, but much less convenient tunable sources in the 5- and 11-μm ranges, respectively [19].

In the ultraviolet, rare gas, and rare-gas–halide excimer lasers show gain over relatively broad regions scattered from 1400 to 3500 Å [20]. Narrow-band operation has been demonstrated in a few cases with cavity structures similar to those in Fig. 1.3. The narrowest linewidths to date have been obtained by employing the excimer medium as an amplifier for frequencies produced as the harmonics of a conventional dye laser.

Techniques for shifting the outputs produced by convenient tunable sources to new frequency regions will be reviewed in the final chapter of this monograph. Many of these techniques employ the same resonant nonlinear processes that have proved useful in nonlinear laser spectroscopy.

1.3 A CATALOG OF NONLINEAR PHENOMENA

In classical absorption, scattering or fluorescence spectroscopy, the intensity of the light reaching the detector is always linearly proportional to the intensity of the incident radiation. In nonlinear spectroscopy, the interesting component of the intensity at the detector varies as some product of incident intensities. There is often a linear component to this intensity that can be orders of magnitude larger than the nonlinear part. These linear background effects will be ignored, or treated as noise sources in what follows.

It is convenient to define a nonlinear spectroscopy signal amplitude E_s which is radiated by the sample as the result of the process being studied. The resulting change in intensity at the detector can be calculated from (1.1.7), but in nonlinear spectroscopy E_s should be regarded as the fundamental quantity. There is no loss of generality; an absorption experiment, for example, can be regarded as an experiment in which a material radiates a signal amplitude which interferes destructively with the incident radiation to reduce the transmitted intensity. Whatever is transmitted through the sample or radiated by the sample as the result of other processes will be described by other amplitudes and intensities.

1.3.1 Saturated Absorption

Consider a medium in which each atom can absorb only one definite frequency, but in which different atoms can absorb different frequencies. Such a situation can result, for example, from the Doppler effect which allows atoms moving away from a light source to absorb a frequency below that absorbed by atoms moving toward the source. The frequencies, however, are the same in the rest frame of each atom. The absorption lines of such a medium are termed *inhomogeneously* broadened, because different atoms are responsible for different portions of the absorption line. If any frequency within the absorption band could excite every atom with the same probability, the line would be *homogeneously* broadened. Such a situation can occur in an atomic beam where the atomic velocities are perpendicular to the direction of propagation of the light [21].

Having absorbed a quantum of radiation, an atom requires a certain average length of time to return to the initial state. During this time period, that atom can no longer absorb light. If the radiation is intense enough, the absorption can be detectably reduced or "saturated." If, in addition, the radiation is narrow band, only atoms that can absorb the radiation frequency are bleached. A "hole" centered at the radiation frequency is burned in the frequency distribution of ground state population. A corresponding peak appears in the excited state population distribution. Such a situation is diagrammed in Fig. 1.6. The depth of this hole depends upon the ratio of the rate of excitation to the repopulation rate of the lower level. Since the rate of excitation is proportional to the incident intensity, the change in the absorption constant κ is also proportional to the pump intensity: $\Delta\kappa = -\kappa' I_1$. The amplitude of the transmitted wave is proportional to the incident amplitude as well as the population difference at the incident frequency. Saturation thus results in an additional transmitted amplitude—the signal amplitude—proportional to the product of pump intensity and probe amplitude,

$$E_t = E_2 \exp(-\kappa l + \kappa' I_1 l) = E_2 \exp(-\kappa l) + \kappa' l E_2 I_1 \exp(-\kappa l)$$
$$= E_2 \exp(-\kappa l) + E_s, \qquad (1.3.1)$$

where E_2 and E_t are the incident and transmitted probe amplitudes, κ the unsaturated absorption constant, l the sample length, I_1 the pump intensity, and the second and third equal signs result from the assumption that the saturation effect is small. Clearly the signal amplitude is proportional to a cubic product of amplitudes, and a saturated absorption line shape function $\kappa'(\omega_1, \omega_2)$ that depends on two frequencies.

$$E_s = \kappa'(\omega_1, \omega_2) l E_2 I_1 \exp(-\kappa l) = (nc/4\pi)\kappa'(\omega_1, \omega_2) l E_2 |E_1|^2 \exp(-\kappa l). \quad (1.3.2)$$

If one laser (the pump laser) is used to burn a hole in an inhomogeneously broadened absorption line, a second (probe) laser scanned across the absorption will detect a profile similar to Fig. 1.6a—an absorption line with a hole in it. Chopping the pump laser or employing polarization techniques can separate the effect of the hole from that of the rest of the line, yielding the narrow profile in Fig. 1.6b. Equation (1.3.2) describes this effect when I_1 and ω_1 denote the pump intensity and frequency and E_2 and ω_2 the

1.3 A Catalog of Nonlinear Phenomena

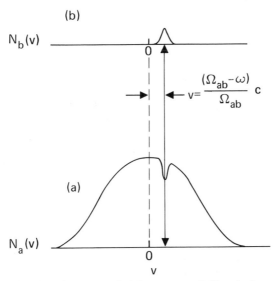

Fig. 1.6 The hole-burning phenomenon in inhomogeneously Doppler-broadened transitions that make saturation spectroscopy possible. A laser at frequency ω depopulates the portion of the ground-state population distribution that is in resonance, while creating an anomalous peak in the excited-state population distribution. The absorption profile thus has a lineshape similar to that shown in (a). Electronic signal processing methods can extract the perturbation with lineshape similar to that shown in (b). The widths of these anomalies are related to the homogeneous linewidth and transverse relaxation time as described in Chapters 2 and 3.

probe amplitude and frequency. The profile in Fig. 1.6b reflects the frequency dependence of κ' which will be related to the nonlinear susceptibility in Chapter 3. If the laser linewidth is small enough, the linewidth in Fig. 1.6b reflects the homogeneous linewidth of the transition.

It is not necessary to employ two different laser frequencies if the inhomogeneous broadening results from the Doppler effect of atoms in a vapor. The saturated absorption signal amplitude in (1.3.2) results from atoms which interact simultaneously with the pump and probe beams. If two beams of frequency ω counterpropagate through a vapor of moving atoms with resonant frequency Ω, atoms moving along the propagation axis with velocity component v absorb one beam when $\Omega = \omega(1 + v/c)$ and the other beam when $\Omega = \omega(1 - v/c)$. Only atoms with $v = 0$ can interact with both beams to produce a saturated absorption signal. The atomic velocities thus automatically produce the frequency difference necessary for saturation spectroscopy. Equivalently, one laser produces a "beam" of saturated atoms with definite velocity which then interact resonantly with the other beam only when that velocity is zero.

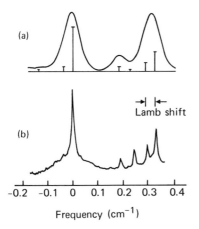

Fig. 1.7 The Balmer α line of atomic deuterium: (a) the emission spectrum; (b) an early saturation absorption spectrum. A narrow splitting due to the quantum electrodynamic Lamb shift is indicated [22].

In Fig. 1.7 the linear absorption profile of the deuterium red Balmer line is compared with a saturated absorption spectrum obtained in the early days of tunable lasers. The increased resolution obtained by suppressing the Doppler width is evident. Term values can thus be determined with increased accuracy, fine and hyperfine structures uncovered, and the effects of collisions, external fields, and other perturbations can be determined. Saturated absorption experiments have resulted in improved values for such fundamental constants as the Rydberg, and may yield improved definitions of the meter, second, and speed of light [22].

Polarization spectroscopy is an important variation of saturation spectroscopy useful when the coupled energy levels are degenerate as in Fig. 1.8 or when the absorbing centers belong to distinguishable orientational classes. In polarization spectroscopy, the pump wave is polarized in such a way that it excites only one of a pair of transitions with equal frequency. As in the previous case, a hole is burned in the ground-state population distribution, while a peak appears in the upper-state population distribution. These excited atoms, however, can decay into either of the two degenerate lower levels. If they decay to the one that can be excited by the pump, the hole is merely filled in a bit. If they decay to the other, a long-lived peak appears in the population distribution.

Light polarized in the same state as the pump beam encounters an absorption line with a hole in it as in the former saturation spectroscopy case. Light polarized orthogonally to the pump encounters an absorption line with a narrow peak at the same frequency as the pump. (In some cases, this peak is absent and the absorption profile for the orthogonal state is unperturbed.) A beam polarized with components in both polarization states will have its polarization state altered by the dichroism induced by

1.3 A Catalog of Nonlinear Phenomena

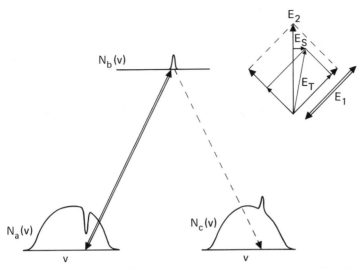

Fig. 1.8 The essence of polarization spectroscopy. A strong pump wave with polarization E_1 couples states $|a\rangle$ and $|b\rangle$, altering their velocity distributions as in Fig. 1.6. An orthogonally polarized field would measure the population difference between levels $|b\rangle$ and $|c\rangle$. The actual probe beam E_2 has components both parallel and perpendicular to E_1 and the absorption and dispersion of the two components will be unequal. The transmitted polarization E_T thus deviates from the direction of E_2. A polarizer selects the component (E_s) perpendicular to E_2 which is detected as the polarization spectroscopy signal amplitude.

the pump beam. This is illustrated in Fig. 1.8 for the case of pump and probe beams linearly polarized at 45° to one another and exactly on resonance. Exquisitely sensitive optical methods exist for detecting such changes in polarization state. In polarization spectroscopy, the signal amplitude is the component of the transmitted probe beam orthogonal to the initial probe polarization. In Section 3.2, we derive an expression for this quantity similar to Eq. (1.3.1) but proportional to the difference in absorption constants and refractive indices for the two components of the probe beam.

A polarization spectroscopy trace in hydrogen similar to the saturation spectroscopy traces in Fig. 1.7 is shown in Fig. 3.22. The sensitivity of polarization spectroscopy techniques can be thousands of times better than saturation techniques. A detailed treatment of sensitivity appears in Chapter 4.

1.3.2 Coherent Raman Effects

Spontaneous Raman scattering has been an important spectroscopic tool since 1928, but the scattered intensities have always been disconcertingly weak. The coherent Raman spectroscopy techniques yield the same information as spontaneous scattering, but with signal levels nine orders of magnitude larger [23]. This enormous improvement results from the fact that laser fields at two different frequencies can force molecules to vibrate at the frequency difference. The vibrating molecules then modulate the dielectric constant, altering the frequency, intensity, or polarization of the incident beam to produce a coherent Raman signal amplitude.

The essence of the process can be derived from Eqs. (1.1.11) and (1.1.12). When the incident field contains two frequency components,

$$E(\mathbf{r}, t) = E_1 \cos(\omega_1 t - \mathbf{k}_1 \cdot \mathbf{r}) + E_2 \cos(\omega_2 t - \mathbf{k}_2 \cdot \mathbf{r}), \quad (1.3.3)$$

these equations imply a force on the generalized coordinate Q that describes molecular vibrations,

$$F_Q = \tfrac{1}{4} \partial \alpha / \partial Q \, E_1 E_2^* \cos[(\omega_1 - \omega_2)t]. \quad (1.3.4)$$

(By convention $\omega_1 > \omega_2$.) If the equation of motion for Q is that of a damped harmonic oscillator with frequency Ω_Q, one can use Newton's second law to calculate the response:

$$\ddot{Q} + \Gamma_Q \dot{Q} + \Omega_Q Q = F_Q. \quad (1.3.5)$$

When $\omega_1 - \omega_2 = \Omega_Q$, Q is driven resonantly with a large amplitude. The oscillating term in the polarizability due to the coherently driven molecular coordinate mixes with an incident frequency giving rise to a dielectric polarization density

$$\mathbf{P}_Q = (\partial \alpha / \partial Q) Q \mathbf{E} \quad (1.3.6)$$

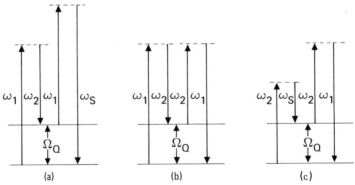

Fig. 1.9 The magnitudes of the photon energies and level splittings in coherent Raman spectroscopy. Such diagrams should not be interpreted as depicting a time-ordered process or the dynamics of population changes. (a) The CARS process, (b) the process responsible for stimulated Raman spectroscopy (SRS) as well as the Raman induced Kerr effect (TRIKE), and (c) coherent Stokes Raman spectroscopy (CSRS).

1.3 A Catalog of Nonlinear Phenomena

which in turn radiates a variety of detectable amplitudes. A more complete theoretical description appears in Chapter 2. The magnitudes of the photon energies and level splittings are shown in Fig. 1.9.

The various experimental techniques of coherent Raman spectroscopy detect different amplitudes. Some of the common techniques are diagrammed in Fig. 1.10. In coherent anti-Stokes Raman spectroscopy (CARS), the detected output frequency is $\omega_s = 2\omega_1 - \omega_2$. This field is radiated by a polarization that results from mixing the component of Q oscillating at $\omega_1 - \omega_2$ with the incident field at ω_1.

A wave-vector matching condition

$$|\mathbf{k}_s - 2\mathbf{k}_1 + \mathbf{k}_2| < \pi/l \tag{1.3.7}$$

must be fulfilled to maximize the output. In (1.3.7), l is the sample length. The intensity at the detector is proportional to the absolute square of the amplitude and polarization, and thus to a cubic product of the incident intensities. Strong pulsed lasers are thus desirable, but the generated beam can be separated spectrally and spatially from the inputs.

If the roles of the frequencies are reversed, this type of interaction yields another coherent Raman technique—coherent Stokes Raman spectroscopy (CSRS—pronounced "scissors"). The output is then at $\omega_s = 2\omega_2 - \omega_1$, below the minimum input frequency. In four-wave mixing a third input frequency ω_0 is supplied to mix with the component of Q oscillating at

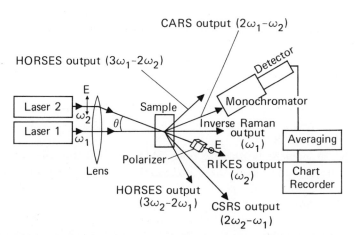

Fig. 1.10 Typical CARS experiment also showing the directions of the output beams for other coherent Raman processes. All such processes require two input frequencies and similar apparatus. The acronym "HORSES" stands for higher order Stokes effect scattering.

$\omega_1 - \omega_2$. The output frequency is then $\omega_s = \omega_0 \pm (\omega_1 - \omega_2)$ and the wave-vector matching condition is

$$|\mathbf{k}_s - \mathbf{k}_0 \pm (\mathbf{k}_1 - \mathbf{k}_2)| < \pi/l. \qquad (1.3.8)$$

When the third input has frequency ω_1, but a different propagation direction from the other beam at that frequency, the wave-vector matching condition can be fulfilled in the "boxcars" geometry of Fig. 1.11b. Such a technique can enhance spatial resolution and probe inhomogeneous samples.

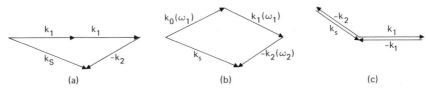

Fig. 1.11 Wave-vector matching diagrams for common coherent Raman processes. In each case, the signal amplitude propagates along $\mathbf{k}_s$. In (a) CARS and (b) 4 WM, the propagation angles must be chosen to fulfill a phase-matching condition, while in (c) SRS, TIRE, and TRIKE, that condition is automatically fulfilled for every angle.

The component of Q oscillating at $\omega_1 - \omega_2$ can mix with the input beam at ω_1 or ω_2 to produce a radiated field at the other input frequency. The amplitude radiated at ω_2 interferes constructively with the incident amplitude at that frequency producing amplification—stimulated Raman gain (SRG). At high enough intensities, such gain can lead to oscillation without an input frequency, but such Raman laser action is useful only for generating new frequency components. In stimulated Raman spectroscopy (SRS), the fractional amplification is very small, and the change in detected intensity with frequency parallels the resonant structure in $\sigma(\omega_1 - \omega_2)$. The amplitude radiated by the medium at ω_1, interferes destructively with the input at that frequency, resulting in stimulated Raman loss (SRL) or the inverse Raman effect (TIRE). When one frequency is produced by a laser with a stable amplitude, these gain and loss effects can be detected with a sensitivity comparable to CARS.

When the incident beams are polarized, neither identically nor orthogonally, the asymmetries in the Raman gain and loss effects can alter the polarization of the beams pretty much as the intensity induced dichroism alters the polarization of the probe in polarization spectroscopy. These phenomena give rise to the Raman induced Kerr effect (TRIKE) and are exploited in Raman induced Kerr effect spectroscopy (RIKES) and its variations. In these techniques, the incident probe beam is approximately linearly polarized, and the pump is either linearly polarized at 45° with respect to the probe or circularly polarized. The detected amplitude appears in a

1.3 A Catalog of Nonlinear Phenomena

direction orthogonal to the initial probe polarization. These techniques allow the suppression of all background signals and yield the same superb sensitivity as polarization spectroscopy.

The relevant wave-vector matching conditions are automatically fulfilled in SRS and RIKES, and thus, the interacting beams can propagate in any convenient direction. When the beams propagate in the same direction, inhomogeneous broadening due to the Doppler effect is minimal. Raman spectra with very high resolution can thus be obtained.

Other imaginatively named wave-mixing geometries and frequency conditions include HORSES (higher-order Stokes effect scattering at $\omega_s = 3\omega_2 - 2\omega_1$, etc.), HORAS (higher-order anti-Stokes scattering at $\omega_s = 3\omega_1 - 2\omega_2$), "Asterisk," and "Submarine." Photoacoustic Raman spectroscopy (PARS) and optoacoustic Raman spectroscopy (OARS) detect the energy deposited in the medium directly. The applicability of these and other techniques will be discussed in Chapter 4.

The fact that the output is a coherent beam of light that can be well separated from scattering, luminescence, and blackbody radiation from the sample allows these coherent Raman techniques to be used to probe adverse environments such as flames, explosions, and plasmas. If the probe laser is broad band rather than single frequency, and a spectrometer is used to disperse the various frequency components, the complete Raman spectrum can be obtained in a single laser shot. This information allows one to identify the species present in the sample as well as the temperatures and occasionally even the quantum state. The high-resolution capability of these coherent Raman techniques makes possible reevaluation of the molecular constants of many molecules with greatly improved precision. In favorable cases, extremely weak Raman modes can be detected, very subtle changes in line shape become apparent and very dilute constituents of mixtures can be identified.

1.3.3 Multiquantum Absorption

Excited states of atoms and molecules can be populated by processes in which more than one photon is absorbed from the radiation field. The rate of such a process depends upon some power of the incident intensity—or equivalently photon flux

$$\Gamma^{(N)} = \hat{\sigma}_N F^N, \tag{1.3.9}$$

where the photon flux $F = I/\hbar\omega$ is measured in quanta per square centimeter per second, and the equivalent multiphoton cross section $\hat{\sigma}_N$ has dimensions $cm^{2N} \sec^{N-1}$ [24].

These multiphoton cross sections can be calculated using Fermi's golden rule. The energy of a state populated by an Nth-order multiphoton process is related to the photon energy by

$$E_f - E_0 = N\hbar\omega. \qquad (1.3.10a)$$

The cross sections for Nth-order multiphoton processes can be greatly enhanced when some smaller number of photons ($M < N$) has energy resonant with an intermediate state, i.e.,

$$E_i - E_0 = M\hbar\omega. \qquad (1.3.10b)$$

The probability for such M-photon-resonant Nth-order nonlinear processes can be fully comparable to the probability for single quantum transitions. The change in transmitted intensity that results from multiquantum absorption is, however, generally too small to detect directly. A variety of sophisticated techniques based upon the detection of fluorescence resulting from the decay of the excited state, of ions, electrons, or neutral species formed in the decay have been developed to detect multiquantum absorption events.

Multiquantum absorption also gives rise to resonances in the nonlinear susceptibility tensors discussed in Sections 2.5 and 2.8. The resulting enhancement in the intensities of waves generated by nonlinear mixing in the medium can be detected and employed to maximize the efficiency of the wave mixing process. The ultraviolet radiation generated in resonant sum frequency generation processes of this sort has proved useful in studying highly excited states.

Multiphoton absorption allows highly excited states of atoms to be studied with convenient visible radiation. Alternatively, when powerful infrared lasers are employed, such highly excited vibrational states of molecules are produced that the molecules can dissociate into constituent radicals and atoms. Occasionally, this process is isotope specific.

When an even number of photons are absorbed, the excited state has the same parity as the initial state. Such transitions are forbidden for single quantum processes. Thus, multiphoton absorption may be the only convenient technique for producing atoms in such levels.

Geometries also exist in which no momentum is transferred from the radiation to the absorbing species. In such geometries,

$$\sum_i \mathbf{k}_i = 0. \qquad (1.3.11)$$

Energy level diagrams and wave-vector matching conditions for three such processes are shown in Fig. 1.12. Under these conditions, Doppler broadening from the thermal motion of the absorbing atoms is suppressed. This

1.3 A Catalog of Nonlinear Phenomena

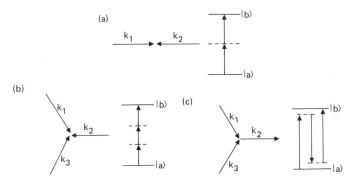

Fig. 1.12 Wave-vector summing and energy level diagrams for Doppler-free two- and three-photon absorption. When an atom undergoing such a transition extracts no momentum from the radiation field, thermal motion does not introduce Doppler broadening.

can be readily seen in Fig. 1.12a in which an atom moving to the right encounters a leftward-propagating wave shifted up in frequency by the same amount as the rightward-moving wave is shifted down. A two-photon resonance in which one photon from each beam is absorbed is unshifted and unbroadened by the motion of the atoms [25]. Angular momentum selection rules can sometimes be employed to suppress all multiphoton processes except those for which the Doppler width vanishes. Since many of the states accessed by multiphoton absorption are metastable states with lifetimes from milliseconds to seconds, these Doppler-free multiquantum absorption processes promise extremely high resolution and may lead to useful new definitions of the second, meter, etc.

Multiphoton processes can also promote atoms to states in the continuum; i.e., ground state atoms can be ionized as the result of simultaneously absorbing several quanta of radiation. At low intensities, these processes can be described by the usual perturbation theory formulas. But at the very highest intensities, the operative effect may be electron tunneling through the potential barrier formed by the Coulomb potential of the atomic nucleus and the sloping potential of the external field. In between, a variety of interesting new phenomena have been observed, resulting in part from large optical Stark shifts of the atomic levels. Electrons emerge from the ionized atoms with distinctive distributions of directions characteristic of the process and of the states nearest resonance.

The motivation for developing the multiquantum absorption techniques results from the interest in highly excited states, especially those with the same party as the ground state, the need for high-resolution techniques in the ultraviolet and vacuum-ultraviolet region, and the possibility that these processes may lead to useful isotope separation techniques or generally

1.3.4 Optical Coherent Transients

When a light wave is absorbed by a homogeneously broadened transition, the observed diminution of transmitted intensity results from a destructive interference between the wave radiated by the sample and the incident beam. If the incident beam is suddenly extinguished, the wave radiated by the sample continues to be radiated, decaying exponentially in a time T_2. This "transverse" relaxation time is related to the homogeneous linewidth (HWHM) by

$$1/T_2 = \Gamma_2. \qquad (1.3.12)$$

The wave radiated freely by the sample after the exciting light is extinguished is called an optical free induction decay (OFID). It is one of the phenomena of optical coherent transient spectroscopy [9, 26].

Techniques employing these phenomena allow careful measurement of the relaxation processes in the time domain. These time-domain techniques complement the frequency-domain techniques previously discussed.

Another useful coherent transient technique is the photon echo most easily explained in the case of an inhomogeneously broadened transition. When a pulse of light is absorbed by such a transition, the free induction decay radiated by the atoms with different resonant frequencies soon go out of phase with one another, leading to a cancellation in the radiated intensity. Each atom, however, continues to radiate its own wave. A large pulse absorbed at a later time can reverse this dephasing process giving rise to a radiated echo pulse at a time still later. The time at which the echo pulse occurs depends upon the interval between the pulses that induced it. Its amplitude, however, decays exponentially with time with rate $1/T_2$. A detailed calculation of the photon echo will appear in Chapter 6.

An even more remarkable phenomenon is the stimulated echo. In this process, two pulses separated by a short time τ are used to excite an inhomogeneously broadened transition. The Fourier transform of such a two-pulse sequence shows maxima and minima in the frequency spectrum. At the frequency of a maximum, a hole is burned in the absorption, whereas at the frequency of a minimum the population is relatively unaffected (see Fig. 6.5). A later probe pulse can interrogate that system. If this fringe structure in the population persists, the atoms will radiate an echo—the stimulated echo—at a time τ after the interrogating pulse. The amplitude of the stimulated echo decays as the population decays to its initial state. In some favorable cases, stimulated echoes have been detected minutes— even hours—after initially excited.

Another class of coherent transient phenomena involve the so-called Ramsey fringe effect. Atoms excited by an initial pulse continue to oscillate whether or not the resulting OFID is detected. A second incident pulse can either increase the excitation by adding in phase with the atomic oscillation or decrease it by adding out of phase. Thus, the level of excitation following the second pulse depends sinusoidally on the product of the time between the pulses and the detuning from resonance. In frequency space, the width of these fringes depends only on the separation between the pulses, but the amplitude of the fringes depends upon various decay rates [27, 28]. Incredibly narrow fringes have been obtained when the Ramsey fringe technique was used to suppress the transit-time broadening of a nonlinear resonance.

Several books and review articles have been written about these coherent transient phenomena. Only the briefest review of the most useful experimental techniques will appear in Chapters 2 and 6.

1.4 LINEAR SPECTROSCOPY WITH NONLINEAR SOURCES

As useful as the dye lasers and other tunable optical sources can be, they do not span the entire electromagnetic spectrum. Their range, however, can be extended using some of the resonant nonlinear optical processes that are also employed in spectroscopy. In particular, the stimulated Raman effect can be used to shift the frequency of a tunable or broadband coherent source into the 2- to 3-μm region of the infrared. Organic molecules have characteristic "fingerprint" spectra in this region. The attenuated and transmitted beam can be upshifted in frequency using various four-wave-mixing phenomena related to coherent anti-Stokes Raman scattering. The outcome of these manipulations is to shift the characteristic absorption spectra in the 2- to 3-μm region into the visible where it can be photographed or recorded by an optical multichannel analyzer. This is the basis of the infrared spectrophotography technique pioneered by Sorokin and Bethune [29]. Because the entire spectrum can be recorded with a single nanoseconds-long laser pulse, this technique can be used to probe explosions, other rapid chemical reactions, and to identify short-lived chemical species.

The stimulated Raman effect and coherent anti-Stokes Raman spectroscopy can be used in a different way to generate radiation in the ultraviolet and vacuum-ultraviolet regions. When a powerful laser beam is loosely focused into a 0.5-m-long cell of high-pressure hydrogen gas, stimulated

Raman oscillation produces one or two orders of Stokes shifted radiation. These Stokes beams then mix with the incident frequency to produce anti-Stokes beams, which mix resonantly with one another to produce higher-order anti-Stokes frequencies. As the laser beam is tuned, so is the frequency of the anti-Stokes radiation. Useful amounts of power—fractions of a millijoule—have been generated at wavelengths down to 1900 Å by this technique.

Even so, sum frequency generation in resonant atomic vapors continues to be useful to generate wavelengths in the ultraviolet and vacuum-ultraviolet regimes. The responsible processes are closely related to multiquantum absorption and to coherent anti-Stokes Raman spectroscopy. Incoherent anti-Stokes radiation has already been generated in the 60-Å region by scattering a strong incident beam from atoms of helium excited by electron impact into the metastable $2s$ state. As crude as they may be, these nonlinear sources produce more power than the sources conventionally used in ultraviolet and vacuum-ultraviolet spectroscopy. Thus, nonlinear laser spectroscopy impacts even the classical areas of UV and IR absorption spectroscopy. In time, perhaps all of spectroscopy may be done with lasers. Some of the advantages and disadvantages of such an approach will be revealed in Chapter 7.

1.5 LASER SPECTROSCOPY LITERATURE

Nonlinear laser spectroscopy is a dynamic and rapidly changing field. By the time this volume appears, new developments may have made large sections obsolete. More up-to-date information will appear in technical journals and the proceedings of conferences. To understand the terminology of these sources, however, one should be familiar with the special terminology of the laser and quantum electronics community.

This book is about nonlinear laser spectroscopy; there are other excellent textbooks on lasers and quantum electronics which provide useful background in areas outside laser spectroscopy. Among them are A. Yariv, "Quantum Electronics," 2nd ed. Wiley, New York, 1975; M. Sargent III, M. O. Scully, and W. E. Lamb, "Laser Physics," Addison-Wesley, Reading, Massachusetts, 1974; R. H. Pantell and H. E. Puthoff, "Fundamentals of Quantum Electronics," Wiley, New York, 1969; and A. E. Siegman, "An Introduction to Lasers and Masers," McGraw-Hill, New York, 1968.

An international conference on laser spectroscopy is held in odd-numbered years. The proceedings are published as "Laser Spectroscopy N" by Springer-Verlag, Berlin. In even-numbered years, the International Quantum Electronics Conference reviews developments in the entire laser area.

Summaries of all the papers are published alternately by *The Journal of the Optical Society of America* and the *IEEE Journal of Quantum Electronics*. There are several other series of conferences with published proceedings, but they place less emphasis on spectroscopy.

The laser spectroscopy community publishes in nearly all of the world's scientific journals. The journals with the highest density of papers in this area are *Optics Letters, Applied Physics Letters, Optics Communications, IEEE Journal of Quantum Electronics, Physical Review Letters, The Journal of the Optical Society of America, Physical Review A* and *B, Journal of Chemical Physics, JETP,* and *Applied Physics*. The *Journal of Quantum Electronics* presents timely review articles and special issues. Longer review articles on mature fields appear in *Progress in Quantum Electronics*. Each year, these articles are collected into a volume published by Pergamon Press, Oxford.

REFERENCES

1. H. A. Lorentz, "The Theory of Electrons." Teubner, Leipzig, 1909.
2. L. I. Schiff, "Quantum Mechanics," 3rd ed. McGraw-Hill, New York, 1968.
3. R. L. Kronig, *J. Opt. Soc. Amer.* **12**, 547 (1926); H. D. Kramers, *Atti. Congr. Int. Fis.* **2**, 545 (1927).
4. O. Howarth, "Theory of Spectroscopy." Halsted, New York, 1973.
5. G. Herzberg, "Infrared and Raman Spectra." Van Nostrand–Reinhold, Princeton, New Jersey, 1945; also D. A. Long, "Raman Spectroscopy." McGraw-Hill, New York, 1977.
6. H. W. Schrotter and H. W. Klockner, Raman scattering cross sections in gases and liquids, *in* "Raman spectroscopy of Gases and Liquids" (A. Weber, ed.) (Topics in Current Physics). Springer-Verlag, Berlin, 1979.
7. A. Abragam, "Principles of Nuclear Magnetism." Oxford Univ. Press, London and New York, 1961.
8. C. H. Townes and A. L. Schawlow, "Microwave Spectroscopy." Dover, New York, 1975.
9. L. Allen and J. H. Eberly, "Optical Resonance and Two-Level Atoms." Wiley (Interscience), New York, 1975.
10. N. Bloembergen, "Nonlinear Optics." Benjamin, New York, 1965, and references therein.
11. F. P. Schäfer, ed., "Dye Lasers" (Springer Series in Applied Physics 1). Springer-Verlag, Berlin, 1977.
12. A. L. Schawlow and C. H. Townes, Infrared and optical lasers, *Phys. Rev.* **112**, 1940–1949 (1958).
13. M. G. Littman and H. J. Metcalf, A spectrally narrow pulsed dye laser without beam expander, *Appl. Opt.* **17**, 2224 (1978); G. K. Klauminer, New high-performance short cavity laser design, *IEEE J. Quant. Electon.* **QE-13**, 92D–93D (September 1977).
14. F. Trehin, F. Biraben, B. Cagnac, and G. Grynberg, Flashlamp pumped tunable dye laser of ultranarrow bandwidth, *Opt. Commun.* **31**, 76–80 (1979).
15. P. Drell and S. Chu, Megawatt dye laser oscillator-amplifier system for high resolution spectroscopy, *Opt. Commun.* **28**, 343–348 (1979).
16. Linn F. Mollenaur, A synopsis of color center lasers and empirical data on laser-active color centers, *in* "CRC Laser Handbook." Chem. Rubber Publi., Cleveland (to be published).

17. R. L. Byer, Parametric oscillators, *in* "Tunable Lasers and Applications" (A. Mooradian, J. Jaeger, and P. Stokseth, eds) (Springer Series in Optical Sciences 3), pp. 70–80. Springer-Verlag, Berlin 1976; R. L. Byer, Optical parametric oscillators, *in* "Quantum Electronics" (H. Rabin and C. L. Tang, eds.), Vol. IB, pp. 587–702. Academic Press, New York, 1975.
18. D. F. Bulter and T. C. Harman, Long wavelength infrared $Pb_{1-x}Sn_xTe$ diode lasers, *Appl. Phys. Lett.* **12**, 347 (1968).
19. C. K. N. Patel and E. D. Shaw, *Phys. Rev. B* **3**, 1279 (1971); A. Mooradian, S. R. J. Brueck, and F. A. Blum, *Appl. Phys. Lett.* **17**, 481 (1971).
20. *IEEE J. Quant. Electron. Spec. Issue Excimer Lasers* **QE-15**, No. 5, 265–398 (1979).
21. A. Yariv, "Quantum Electronics," 2nd ed., p. 165. Wiley, New York, 1975.
22. T. W. Hänsch, Nonlinear high resolution spectroscopy of atoms and molecules, *in* "Nonlinear Spectroscopy" (Proc. Int. School Phys., Enrico Fermi, Course 64) (N. Bloembergen, ed.). North-Holland Publi. Amsterdam, 1977.
23. M. D. Levenson and J. J. Song, Coherent Raman spectroscopy, *in* "Coherent Nonlinear Optics" (M. S. Field and V. S. Letokhov, eds.) (Topics on Current Physics 27). Springer-Verlag, Berlin, 1980.
24. P. Lambropoulos, Topics on multiphoton processes in atoms, *Advan. Mol. Phys.* **12**, 87 (1976).
25. N. Bloembergen and M. D. Levenson, Doppler-free two photon absorption spectroscopy, *in* "High Resolution Laser Spectroscopy" (K. Shimoda, ed.) (Topics in Applied Physics 13). Springer-Verlag, Berlin, 1976.
26. R. G. Brewer, Coherent optical spectroscopy, *in* "Nonlinear Spectroscopy" (Proc. Int. School Phys., Enrico Fermi, Course 64) (N. Bloembergen, ed.). North-Holland Publ., Amsterdam, 1977.
27. M. M. Salour, Ultra-high-resolution two-photon spectroscopy in atomic and molecular vapors, *Ann. Phys.* **111**, 365–503 (1978).
28. V. P. Chebotayev, Coherence in high resolution spectroscopy, *in* "Coherent Nonlinear Optics" (M. S. Feld and V. S. Letokhov, eds.) (Topics in Current Physics 21). Springer-Verlag, Berlin, 1980.
29. P. M. Avouris, D. S. Bethune, J. R. Lankard, J. A. Ors, and P. P. Sorokin, Time resolved infrared spectrophotography: Study of laser initiated explosions in NH_3, *J. Chem. Phys.* **74**, 2304–2312 (1981).

Chapter 2
THEORY

2.1 THE DENSITY MATRIX FOR A TWO-LEVEL SYSTEM

The simplest quantum-mechanical system consists of an isolated entity with energy eigenstates $|b\rangle$ and $|a\rangle$ having energies E_b and E_a ($E_b > E_a$), respectively. If the entity is in the ground state $|a\rangle$, the wave function describing the system is

$$|\Psi\rangle = e^{-iE_a t/\hbar}|a\rangle \tag{2.1.1}$$

and the energy is

$$\langle\Psi|i\hbar\frac{\partial}{\partial t}|\Psi\rangle = E_a. \tag{2.1.2}$$

Similarly, if the entity is known to be in the excited state, the wave function and energy are

$$|\Psi\rangle = e^{-iE_b t/\hbar}|b\rangle \quad \text{and} \quad \langle\Psi|i\hbar\frac{\partial}{\partial t}|\Psi\rangle = E_b. \tag{2.1.3}$$

The most general wave function for this system, however, describes a coherent superposition state

$$|\Psi\rangle = a_\Psi e^{-iE_a t/\hbar}|a\rangle + b_\Psi e^{-i\phi_\Psi}e^{-iE_b t/\hbar}|b\rangle \tag{2.1.4}$$

where a_Ψ, b_Ψ, and ϕ_Ψ are real numbers, $a_\Psi^2 + b_\Psi^2 = 1$ and an arbitrary overall phase factor has been set to unity. The wave function in Eq. (2.1.4) describes a system which has probability a_Ψ^2 of being found in the ground state $|a\rangle$ and probability b_Ψ^2 of being in the excited eigenstate $|b\rangle$. All measurable properties of a two-level system in this state can be calculated by taking the expectation value of the operator describing the property. For example, the expectation value of the energy is

$$\langle \Psi | i\hbar \frac{\partial}{\partial t} | \Psi \rangle = a_\Psi^2 E_a + b_\Psi^2 E_b. \tag{2.1.5}$$

The density matrix provides an alternative representation of the quantum mechanics of a two-level system.[1] In terms of the wave function Ψ, the density matrix for an isolated entity is defined as

$$\rho = |\Psi\rangle\langle\Psi| \tag{2.1.6}$$

in the representation where $|a\rangle = \binom{1}{0}$ and $|b\rangle = \binom{0}{1}$; the density matrix equivalent to the general wave function of (2.1.4) is

$$\rho = \begin{pmatrix} a_\Psi e^{-iE_a t/\hbar} \\ b_\Psi e^{-i(E_b t/\hbar + \phi_\Psi)} \end{pmatrix} \begin{pmatrix} a_\Psi e^{+iE_a t/\hbar} & b_\Psi e^{+i(E_b t/\hbar + \phi_\Psi)} \end{pmatrix}$$

$$= \begin{pmatrix} a_\Psi^2 & a_\Psi b_\Psi e^{+i\phi_\Psi} e^{-i(E_a - E_b)t/\hbar} \\ a_\Psi b_\Psi e^{-i\phi_\Psi} e^{i(E_a - E_b)t/\hbar} & b_\Psi^2 \end{pmatrix}. \tag{2.1.7}$$

All measurable properties of the system can be calculated using the density matrix; the expectation value for the operator $\tilde{\mathcal{O}}$ is

$$\langle \tilde{\mathcal{O}} \rangle = \text{Tr}(\rho \tilde{\mathcal{O}}), \tag{2.1.8}$$

where the matrix operator Tr denotes the trace of the quantity in parentheses. Thus, the expectation value of the energy can be calculated using the fact that the energy operator in the present representation is

$$i\hbar \frac{\partial}{\partial t} \equiv \begin{pmatrix} E_a & 0 \\ 0 & E_b \end{pmatrix} \tag{2.1.9}$$

and

$$\text{Tr}\left(i\hbar \frac{\partial}{\partial t} \rho\right) = a_\Psi^2 E_a + b_\Psi^2 E_b. \tag{2.1.10}$$

The density matrix obeys the Liouville variant of the Schroedinger equation

$$i\hbar\rho = [\mathcal{H}, \rho] \tag{2.1.11}$$

---

[1] A number of texts on quantum mechanics derive the density matrix and its properties with full rigor, particularly recommended is that by Sargent et al. [1].

2.1 The Density Matrix for a Two-Level System

where $\tilde{\mathscr{H}}$ is the Hamiltonian of the system. The diagonal elements of ρ represent the probabilities for finding the system in the two basis states (in this representation, the two energy eigenstates). The off-diagonal elements represent the *coherence* intrinsic to a superposition state. The underlying concepts of coherence and phase are central to nonlinear laser spectroscopy.

The real advantage of the density matrix is that it can correctly describe the observable properties of an *ensemble* of quantum-mechanical systems. For an ensemble, the correct definition of the density matrix is

$$\rho = \sum_{\Psi} p_{\Psi} |\Psi\rangle\langle\Psi|, \tag{2.1.12}$$

where p_{Ψ} is the probability of finding the state $|\Psi\rangle$ in the ensemble, and the sum is over all possible quantum states.

If every system in the ensemble were known to be in the state $|\Psi\rangle$ of (2.1.4), $p_{\Psi} = 1$ for that state, and the density matrix in (2.1.7) would describe the ensemble. Very special preparation is required to achieve such a situation. More generally, the ensemble contains a distribution of quantum states. One illustrative distribution is the case where the probabilities for finding the system in the two eigenstates are the same for each state in the distribution, but the *phases* of the coherent superpositions are random. In this case, the wave functions in the ensemble are described by (2.1.4) with a_{Ψ} and b_{Ψ} constant, but with ϕ_{Ψ} randomly distributed between 0 and 2π. The probability of finding a state with a value of ϕ_{Ψ} between ϕ and $\phi + d\phi$ is thus

$$p_{\Psi} = \begin{cases} d\phi/2\pi, & 0 < \phi < 2\pi, \\ 0, & \text{otherwise}, \end{cases} \tag{2.1.13}$$

and the resulting density matrix is

$$\begin{aligned}\rho &= \sum_{\Psi} p_{\Psi} |\Psi\rangle\langle\Psi| \\ &= \int_0^{2\pi} \frac{d\phi}{2\pi} \begin{pmatrix} a_{\Psi}^2 & a_{\Psi} b_{\Psi} e^{+i\phi} e^{-i(E_a - E_b)t/\hbar} \\ a_{\Psi} b_{\Psi} e^{-i\phi} e^{i(E_a - E_b)t/\hbar} & b_{\Psi}^2 \end{pmatrix} \\ &= \begin{pmatrix} a_{\Psi}^2 & 0 \\ 0 & b_{\Psi}^2 \end{pmatrix},\end{aligned} \tag{2.1.14}$$

where we have used $\int_0^{2\pi} d\phi\, e^{i\phi} = 0$. The off-diagonal elements are zero, indicating an absence of coherence in this ensemble. Such a situation cannot be described by any wave function for the ensemble.

Excited but incoherent ensembles of this sort commonly result when the histories of different members of the ensemble are unrelated. Cases intermediate between the complete coherence of (2.1.7) and (2.1.14) are also significant.

The density matrix for an ensemble also obeys Liouville's equation (2.1.11) and the ensemble-averaged expectation value for an operator can be found using (2.1.8). The contributions of each member of the ensemble are summed together by this formalism, and the total can be obtained by multiplying the expectation value by the number of members of the ensemble or by the number density.

2.2 THE INTERACTIONS AND THE HAMILTONIAN

It is convenient to separate the total Hamiltonian of the system into several contributions

$$\tilde{\mathcal{H}} = \tilde{\mathcal{H}}_0 + \tilde{\mathcal{H}}_1(t) + \tilde{\mathcal{H}}_R, \tag{2.2.1}$$

where

$$\tilde{\mathcal{H}}_0 = \begin{bmatrix} E_a & 0 \\ 0 & E_b \end{bmatrix} \tag{2.2.2}$$

is the Hamiltonian describing the internal workings of the isolated two-level systems; it specifies the zeroth-order energy eigenstates.

The interactions between the two-level system and an applied field are described by the interaction Hamiltonian $\tilde{\mathcal{H}}_1(t)$. It is this contribution that is responsible for transitions among the zeroth-order energy eigenstates. In the simplest and most common case, the matrix describing the interactions has zeros along the diagonal and a sinusoidal time dependence. If the states $|a\rangle$ and $|b\rangle$ are coupled by an electric dipole transition, the interaction Hamiltonian that corresponds to the incident wave

$$E(t) = E_0(t)\cos(\omega t + \phi) \tag{2.2.3}$$

is

$$\tilde{\mathcal{H}}_1(t) = -\tilde{\mu} \cdot E(t) = \begin{bmatrix} 0 & -\mu_{ab} \cdot E_0(t) \\ -\mu_{ba} \cdot E_0(t) & 0 \end{bmatrix} \cos(\omega t + \phi). \tag{2.2.4}$$

Similar operators can be derived for multiquantum transitions using the methods of Wilson-Gordon *et al.*, Section 2.5, this volume.

2.3 RELAXATION

The relaxation Hamiltonian $\tilde{\mathcal{H}}_R$ describes all of the processes that return the ensemble to thermal equilibrium [1–6]. The most important of these processes are spontaneous emission, collisions, and—in molecules—coupling between rotational, vibrational, and electronic excitations. Because of

2.3 Relaxation

the complexity of these phenomena, they are usually dealt with in a quasi-phenomenological manner.[2]

Processes similar to spontaneous emission that result in decays from state $|b\rangle$ to state $|a\rangle$ can be described by the relaxation operator matrix elements

$$(i\hbar)^{-1}[\tilde{\mathscr{H}}_R, \rho]_{bb} = -\rho_{bb}/T_b \tag{2.3.1}$$

$$(i\hbar)^{-1}[\tilde{\mathscr{H}}_R, \rho]_{aa} = \rho_{bb}/T_b = (1 - \rho_{aa})/T_b. \tag{2.3.2}$$

The quantity T_b is thus the lifetime of the excited state. The lifetime of state $|a\rangle$ is assumed infinite. The second equal sign in (2.3.2) results from the conservation of particles condition

$$\rho_{aa} + \rho_{bb} = a_\psi^2 + b_\psi^2 = 1. \tag{2.3.3}$$

The off-diagonal elements of the density matrix in this two-level atom approximation also decay toward an equilibrium, but in a different period—conventionally called T_2, the "transverse" relaxation time. The appropriate matrix elements of the relaxation operator are

$$(i\hbar)^{-1}[\tilde{\mathscr{H}}_R, \rho]_{ab} = -\rho_{ab}/T_2 \quad \text{and} \quad (i\hbar)^{-1}[\tilde{\mathscr{H}}_R, \rho]_{ba} = -\rho_{ba}/T_2. \tag{2.3.4}$$

The transverse relaxation time T_2 is related to the lifetimes of the eigenstates by

$$T_2^{-1} = \tfrac{1}{2}(T_a^{-1} + T_b^{-1}) + \gamma_\phi, \tag{2.3.5}$$

where γ_ϕ is the rate of events that perturb only the phase of the wave functions without inducing decay of eigenstates, and T_a is the lifetime of the lower level. In our two-level atom approximation, $T_2^{-1} = \tfrac{1}{2} T_b^{-1} + \gamma_\phi$; γ_ϕ is called the pure dephasing rate and is attributed mostly to collisions. Intuitively, T_2 is the lifetime of a coherent superposition state, while T_a and T_b are the lifetimes of the energy eigenstates.

More complex treatments of relaxation violate the conservation condition of (2.3.3). Such situations occur naturally whenever the states $|a\rangle$ and $|b\rangle$ can interact with a reservoir of other levels with similar energies. The most common treatment of this type posits thermal equilibrium values for ρ_{aa} and ρ_{bb} and allows each density matrix element to relax toward equilibrium at its own rate:

$$(i\hbar)^{-1}[\tilde{\mathscr{H}}_R, \rho]_{bb} = (\rho_{bb}^e - \rho_{bb})/T_b, \tag{2.3.6}$$

$$(i\hbar)^{-1}[\tilde{\mathscr{H}}_R, \rho]_{aa} = (\rho_{aa}^e - \rho_{aa})/T_a. \tag{2.3.7}$$

When the properties of states $|a\rangle$ and $|b\rangle$ are very similar, one can set $T_a = T_b = T_1$, and T_1 is then termed the "longitudinal" relaxation time. This approximation is widely employed in infrared, microwave and NMR spec-

---

[2] In laser spectroscopy relaxation, models developed for nuclear magnetic resonances have been widely adopted, even when not fully persuasive (see Chapter 1, Ref. [7]).

troscopy. The transverse relaxation time remains an independent quantity given by (2.3.5). In the absence of pure dephasing ($\gamma_\phi = 0$), $T_1 = T_2$ and the dynamics is greatly simplified [6]. In more complex situations, more complex treatments of relaxation can be appropriate.

2.4 THE MASTER EQUATION AND THE VECTOR MODEL

The equations of motion for the elements of the density matrix are termed master equations, and can be written down directly using (2.1.11), (2.2.1), and (2.2.2). It is more convenient, however, to introduce some new notation at this stage. The transition frequency will be denoted $\Omega = (E_b - E_a)/\hbar$ and the strength of the sinusoidal interaction will be parameterized by a generalized Rabi frequency $\chi(t)$. If

$$\tilde{\mathscr{H}}_1(t) = \sum_\omega \begin{bmatrix} 0 & \tilde{\mathscr{H}}_\omega(t) \\ \tilde{\mathscr{H}}_\omega^*(t) & 0 \end{bmatrix} \cos \omega t, \qquad (2.4.1)$$

then

$$\chi_\omega(t) = \tilde{\mathscr{H}}_\omega(t)/2\hbar. \qquad (2.4.2)$$

In terms of these quantities, the master equations become

$$\dot{\rho}_{aa} = \frac{i}{2} \sum_\omega [(\rho_{ba}\chi_\omega(t) - \rho_{ab}\chi_\omega^*(t))(e^{i\omega t} + e^{-i\omega t})] + [\tilde{\mathscr{H}}_R, \rho]_{aa}/i\hbar, \qquad (2.4.3a)$$

$$\dot{\rho}_{bb} = -\frac{i}{2} \sum_\omega [(\rho_{ba}\chi_\omega(t) - \rho_{ab}\chi_\omega^*(t))(e^{i\omega t} + e^{-i\omega t})] + [\tilde{\mathscr{H}}_R, \rho]_{bb}/i\hbar, \qquad (2.4.3b)$$

$$\dot{\rho}_{ab} = \frac{i}{2} (\rho_{bb} - \rho_{aa}) \sum_\omega (\chi_\omega(t)(e^{i\omega t} + e^{i\omega t})) - i\Omega\rho_{ab} + [\tilde{\mathscr{H}}_R, \rho]_{ab}/i\hbar, \qquad (2.4.3c)$$

$$\rho_{ba} = \rho_{ab}^*, \qquad (2.4.3d)$$

where the relaxation operators of (2.3.1) to (2.3.7) can be inserted as necessary. These equations can be solved directly (and in some cases *must* be solved directly), but more insight results from a series of manipulations.

If $\omega \approx \Omega$, the terms on the right of (2.4.3c) proportional to $e^{-i\omega t}$ are much more important in driving the system than those proportional to $e^{i\omega t}$ while the terms that are effective in altering ρ_{aa} and ρ_{bb} oscillate at relatively low frequency. The *rotating-wave approximation* achieves a significant simplification by ignoring the less effective terms. In the rotating-wave approximation, the interaction Hamiltonian becomes

$$\tilde{\mathscr{H}}_1(t) = \sum_\omega \begin{bmatrix} 0 & \mathscr{H}_\omega(t)e^{-i\omega t} \\ \mathscr{H}_\omega^*(t)e^{+i\omega t} & 0 \end{bmatrix} \qquad (2.4.4)$$

2.4 The Master Equation and the Vector Model

which results in the master equations

$$\dot{\rho}_{aa} = \frac{i}{2} \sum_{\omega} [\rho_{ba}\chi_{\omega}(t)e^{-i\omega t} - \rho_{ab}\chi_{\omega}^{*}(t)e^{i\omega t}] + [\tilde{\mathscr{H}}_{R}, \rho]_{aa}/i\hbar, \quad (2.4.5a)$$

$$\dot{\rho}_{bb} = -\frac{i}{2} \sum_{\omega} [\rho_{ba}\chi_{\omega}(t)e^{-i\omega t} - \rho_{ab}\chi_{\omega}^{*}(t)e^{i\omega t}] + [\tilde{\mathscr{H}}_{R}, \rho]_{bb}/i\hbar, \quad (2.4.5b)$$

$$\dot{\rho}_{ab} = \frac{i}{2}(\rho_{bb} - \rho_{aa}) \sum_{\omega} \chi_{\omega}(t)e^{-i\omega t} - i\Omega\rho_{ab} + [\tilde{\mathscr{H}}_{R}, \rho]_{ab}/i\hbar, \quad (2.4.5c)$$

$$\rho_{ba} = \rho_{ab}^{*}. \quad (2.4.5d)$$

The physics omitted in this approximation, however, gives rise to sum frequency generation effects that are potentially useful.

Bloch and later Feynman, Vernon, and Hellwarth contrived an illuminating geometrical picture of this interaction [6]. An abstract three-dimensional vector space defined by the unit vectors $\hat{1}', \hat{2}', \hat{3}'$ is illustrated in Fig. 2.1. The 2×2 density matrix of a two-level system can be mapped

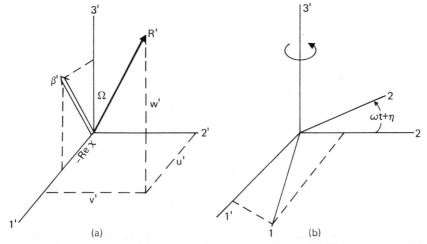

Fig. 2.1 The vector model and the rotating frame. (a) The Bloch vector $\mathbf{R}'$ and the pseudofield vector $\boldsymbol{\beta}'$ defined in the nonrotating reference frame with axes labeled 1', 2', and 3'. When the generalized Rabi frequency χ defined in (2.4.2) is real, the pseudofield vector lies in the 1'-3' plane as shown. More generally, its components are given by (2.4.8). The components of the Bloch vector $\mathbf{R}'$ are illustrated for the general case. (b) The relationship between the nonrotating frame and the rotating frame. The reference frame defined by unit vectors along the 1, 2, and 3 axes rotates with respect to the 1', 2', 3' frame at circular frequency ω. The axis of rotation is the 3' axis which is collinear with the 3 axis of the rotating frame. The Bloch vector in the rotating frame $\mathbf{R}$ is defined by (2.4.11), while the most common convention for the pseudofield vector $\boldsymbol{\beta}$ in the rotating frame is given in (2.4.12). It is occasionally convenient to allow the $\boldsymbol{\beta}$ vector to have a component along $\hat{2}$, in which case (2.4.8) applies with Δ replacing Ω.

into a vector in this space by the Feynman–Vernon–Hellwarth strategem: define the Bloch vector

$$\mathbf{R} = u'\hat{1}' + v'\hat{2}' + w'\hat{3}', \tag{2.4.6}$$

with components

$$u' = \rho_{ab} + \rho_{ba}, \tag{2.4.7a}$$

$$v' = i(\rho_{ba} - \rho_{ab}), \tag{2.4.7b}$$

$$w' = \rho_{bb} - \rho_{aa}. \tag{2.4.7c}$$

At thermal equilibrium, the Bloch vector points downward along the $\hat{3}'$ axis, indicating that most of the two-level systems are in the lower energy eigenstate and that coherence is absent. A second vector—the pseudofield vector—can be defined by

$$\boldsymbol{\beta} = -\operatorname{Re} \chi(t)\hat{1}' - \operatorname{Im} \chi(t)\hat{2}' + \Omega\hat{3}', \tag{2.4.8}$$

where

$$\chi(t) = \sum_\omega \chi_\omega(t)e^{i\omega t} \tag{2.4.9}$$

When $\chi(t) = 0$, the pseudofield is also parallel to the 3' axis, but when the interaction is turned on, the tip of this vector maps out a figure in a plane $\boldsymbol{\beta} \cdot \hat{3}' = \Omega$. If the external field is sinusoidal and the rotating-wave approximation is employed, the tip of the vector traverses a circle of radius χ_ω at angular frequency ω.

The equations of motion for the density matrix can be expressed in terms of $\boldsymbol{\beta}$ and $\mathbf{R}$ as

$$d\mathbf{R}/dt = \boldsymbol{\beta} \times \mathbf{R} + [\text{relaxation terms}]. \tag{2.4.10}$$

If relaxation is neglected for the moment, (2.4.10) indicates that the vector $\mathbf{R}$ precesses around the moving pseudofield $\boldsymbol{\beta}$ just as the angular momentum vector of a magnetized top would process around a variable applied field.

If only one applied frequency is significant, the mad gyrations of these vectors can be simplified by a unitary transformation which sets up a reference frame that rotates around the 3' axis at the frequency of the applied field. The unit vectors of the rotating reference frame are

$$\hat{1} = \cos(\omega t + \eta)\hat{1}' + \sin(\omega t + \eta)\hat{2}', \tag{2.4.11a}$$

$$\hat{2} = -\sin(\omega t + \eta)\hat{1}' + \cos(\omega t + \eta)\hat{2}', \tag{2.4.11b}$$

$$\hat{3} = \hat{3}', \tag{2.4.11c}$$

in terms of those of the fixed frame and the initial phase η of the complex

2.4 The Master Equation and the Vector Model

amplitude $\chi_\omega(t)$. The pseudofield vector is now fixed in the $\hat{1}$-$\hat{3}$ plane

$$\boldsymbol{\beta} = \chi_\omega(t)e^{-i\eta}\hat{1} + \Delta\hat{3} \tag{2.4.12}$$

so long as the phase of $\chi_\omega(t)$ does not change, where $\Delta = \Omega - \omega$ is the detuning from resonance, and the components of the Bloch–Feynman vector become in the rotating frame

$$u = u'\cos\omega t - v'\sin\omega t = 2\,\text{Re}(\rho_{ab}e^{-i\omega t}), \tag{2.4.13a}$$

$$v = u'\sin\omega t + v'\cos\omega t = 2\,\text{Im}(\rho_{ab}e^{-i\omega t}), \tag{2.4.13b}$$

$$w = w' = \rho_{bb} - \rho_{aa}. \tag{2.4.13c}$$

The equation of motion in (2.4.10), however, continues to describe the system. The Bloch–Feynman vector now precesses around a pseudofield vector that varies only with *changes* in the amplitude, frequency, and phase of the driving Hamiltonian. The precession is similar to that of a top in the gravitational field of the earth (see Fig. 2.2). The same result could have been obtained by applying the transformation

$$\tilde{U} = \begin{bmatrix} e^{+i\omega t/2} & 0 \\ 0 & e^{-i\omega t/2} \end{bmatrix} \tag{2.4.14}$$

to the density matrix and Hamiltonian before applying the definitions of (2.4.7) and (2.4.8).

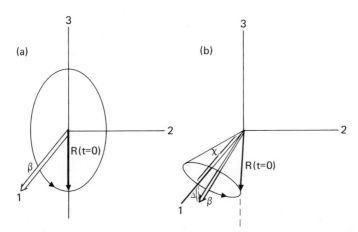

Fig. 2.2 (a) The precession of the Bloch vector around the pseudofield vector in the absence of relaxation. (b) If the fields are suddenly applied to a system in thermal equilibrium at $t = 0$, Eq. (2.4.10) causes the tip of the **R** vector to precess in a circle. The projection of **R** along $\boldsymbol{\beta}$ remains constant as does the magnitude of **R**. In the exact resonance condition illustrated in (a), $\boldsymbol{\beta}$ points along the 1 axis and the tip of **R** traces out a circle in the 2-3 plane. Off resonance as in (b), the circle is smaller and in a plane perpendicular to $\boldsymbol{\beta}$. The line of the vector traces out a cone.

As before, the **R** and **β** vectors are parallel to the 3 axis in the absence of an applied field. If a resonant interaction of amplitude χ_0 is suddenly turned on at $t = 0$, the **β** vector points along the 1 axis,

$$\boldsymbol{\beta} = \chi_0 \hat{1},$$

and the R vector processes around **β** until at time $t = \pi/2\chi_0$ it points along the $\hat{2}$ axis. In this state, the difference in population between states $|a\rangle$ and $|b\rangle$ is zero. At the time $t = \pi/\chi_0$ the **R** vector points upward along the $\hat{3}$ axis, and the population of the two-level system is inverted. Still later at $t = 2\pi/\chi_0$, the **R** vector has returned to its initial state. On balance, no energy has been transferred from the applied fields to the ensemble. A pulse capable of causing the **R** vector to rotate through an angle of 2π in this way is termed a "2π pulse." Such a pulse will propagate unattenuated through an absorbing medium if its length is short enough that relaxation can be ignored. More generally, the quantity

$$\theta = \int_0^t \chi_\omega(t')\,dt', \tag{2.4.15}$$

called the "pulse area" indicates the angle through which the **R** vector rotates when the incident frequency is resonant with the transition [6, p. 78 ff].

If the incident frequency is detuned from exact resonance, and the field applied suddenly, the **R** vector precesses around the pseudofield tracing out a cone with the axis along **β** and cone angle $\tan^{-1}(\chi_\omega/\Delta)$. The precession frequency is $\beta = |\boldsymbol{\beta}| = (\Delta^2 + \chi_\omega^2)^{1/2}$, sometimes called the total Rabi flopping frequency.

The analytical solution of (2.4.10) for the off-resonant case without relaxation is rather complicated [6]. If the initial value of the Bloch–Feynman vector is $\mathbf{R}_0$, and if the generalized Rabi frequency is constant $\chi_\omega(t) = \chi_0$, then the value of R at later times is given by

$$\mathbf{R}(t) = \tilde{G}(t)\mathbf{R}_0, \tag{2.4.16}$$

where the unitary matrix $\tilde{G}(t)$ is given by

$$\tilde{G}(t) = \begin{bmatrix} \dfrac{\chi_0^2 + \Delta^2 \cos\beta t}{\beta^2} & -\dfrac{\Delta}{\beta}\sin\beta t & -\dfrac{\chi_0 \Delta}{\beta^2}(1 - \cos\beta t) \\ \dfrac{\Delta}{\beta}\sin\beta t & \cos\beta t & \dfrac{\chi_0}{\beta}\sin\beta t \\ -\dfrac{\Delta\chi_0}{\beta^2}(1 - \cos\beta t) & -\dfrac{\chi_0}{\beta}\sin\beta t & \dfrac{\Delta^2 + \chi_0^2 \cos\beta t}{\beta^2} \end{bmatrix} \tag{2.4.17}$$

The oscillatory dependence of the populations of the levels coupled by the electromagnetic field implied by Eq. (2.4.17) has an origin similar to

2.4 The Master Equation and the Vector Model

the quantum mechanical oscillation which takes place when two square-well potentials are brought near one another. In that case, a particle initially in one well can tunnel into the other and then back at a frequency characteristic of the coupling between the wells. The true eigenstates of the double-well system have some amplitude for finding the particle in each well, and the symmetric and antisymmetric states are split by the tunneling frequency. The eigenstates of a two-level system coupled strongly by radiation are split by the coupling into two Autler–Townes levels with separation β. The uncoupled upper and lower states can be represented as linear combinations of the Autler–Townes states just as the state with the particle in one well can be represented as a combination of the even and odd parity eigenstates. Equivalent calculations can be made with either representation, but the amplitudes will oscillate when the states employed are not eigenstates of the coupled radiation–matter system. The most popular theory of the coupled system is called the "dressed atom" approach, but its development is beyond the scope of this monograph. [5].

When the time variation in the applied field is slow in comparison to the rate at which $\mathbf{R}$ precesses around β, the $\mathbf{R}$ vector adiabatically follows β as the frequency and intensity is varied. The resulting approximate solution correctly describes the time dependence so long as $d\chi_\omega/dt \ll \Delta^2 + \chi_\omega^2(t)$.

$$u = \frac{\chi_\omega(t)}{(\Delta^2 + \chi_\omega^2(t))^{1/2}}, \qquad (2.4.18a)$$

$$v = \frac{\Delta}{(\Delta^2 + \chi_\omega^2(t))^{3/2}} \frac{d\chi_\omega(t)}{dt} \approx 0, \qquad (2.4.18b)$$

$$w = \frac{\Delta}{(\Delta^2 + \chi_\omega^2(t))^2}. \qquad (2.4.18c)$$

Equations (2.4.18) apply only for time much less than T_2 and T_b; the vector relationships [6] are illustrated in Fig. 2.3.

Relaxation complicates the motions of the $\mathbf{R}$ vector in the rotating frame. In the two-level atom approximation, the equations of motion for the components become

$$\dot{u} + \Delta v + u/T_2 = 0, \qquad (2.4.19a)$$

$$\dot{v} - \Delta u - w\chi_\omega(t) + v/T_2 = 0, \qquad (2.4.19b)$$

$$\dot{w} + \chi_\omega(t)v + (1 + w)/T_b = 0. \qquad (2.4.19c)$$

In the T_1, T_2 approximation, $(w - w^e)$ and T_1 replace $(1 + w)$ and T_b, respectively, in (2.4.19c). In equilibrium $\mathbf{R} = -\hat{3}$ or $w^e\hat{3}$, depending upon the

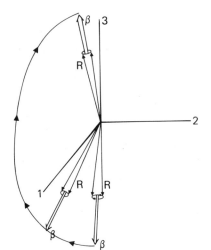

Fig. 2.3 Adiabatic following. When the direction of $\boldsymbol{\beta}$ changes slowly in comparison with the frequency $|\boldsymbol{\beta}|$, the Bloch vector follows the pseudofield maintaining a constant projection along $\boldsymbol{\beta}$. Since both $\mathbf{R}$ and $\boldsymbol{\beta}$ point along the 3 axis when $\chi(t) = 0$ and the system is in equilibrium (i.e., before the light is turned on), the cone angle between $\mathbf{R}$ and $\boldsymbol{\beta}$ is usually very small. The adiabatic following approximation ignores the rapid precession of $\mathbf{R}$ around $\boldsymbol{\beta}$ and assumes that the two are collinear.

approximation. When relaxation is included, the magnitude of the $\mathbf{R}$ vector no longer remains constant. When $\chi(t) = 0$, the projection of $\mathbf{R}$ along the $\hat{3}$ axis relaxes toward its equilibrium value at one rate, the longitudinal relaxation rate, while the components in the $\hat{1}$, $\hat{2}$ plane oscillate and relax toward zero at a different rate, the transverse relaxation rate. The time evolution of the $\mathbf{R}$ vector for various combinations of relaxation times is illustrated in Fig. 2.4.

With the driving fields turned off, the appropriate analytical solutions of (2.4.19) in the reference frame rotating at frequency ω are

$$u(t) = [u_0 \cos \Delta t - v_0 \sin \Delta t] e^{-t/T_2}, \quad (2.4.20a)$$

$$v(t) = [v_0 \cos \Delta t + u_0 \sin \Delta t] e^{-t/T_2}, \quad (2.4.20b)$$

$$w(t) = 1 + [w_0 + 1] e^{-t/T_b}, \quad (2.4.20c)$$

where u_0, v_0, and w_0 are the components of the $\mathbf{R}$ vector at the time when the driving fields are switched off (i.e., $t = 0$).

Torrey gave detailed solutions to (2.4.19) in the T_1, T_2 approximation, and the relatively simple results for certain special cases have been widely quoted [6].

If $\chi_\omega(t) = \chi_0$, a constant, the density matrix evolves toward the steady-state condition illustrated in Fig. 2.5 and in which the components of $\mathbf{R}$ are

$$u(\infty) = -\frac{\Delta\chi_0\{-1\}}{\Delta^2 + T_2^{-2} + \chi_0^2 T_b/T_2}, \quad (2.4.21a)$$

$$v(\infty) = \frac{\{-1\}\chi_0}{T_2} \frac{1}{\Delta^2 + T_2^{-2} + \chi_0^2 T_b/T_2}, \quad (2.4.21b)$$

2.4 The Master Equation and the Vector Model

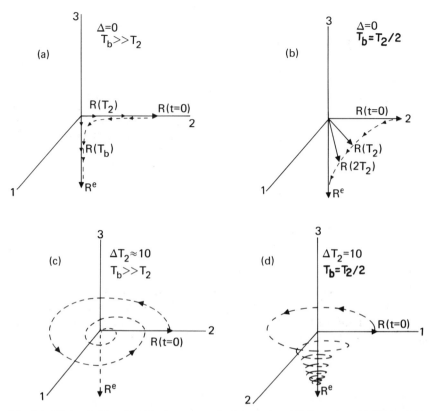

Fig. 2.4 The evolution of the Bloch vector toward thermal equilibrium when the driving fields have been turned off at $t = 0$. (a, b) The rotation frequency is resonant with the transition frequency. (a) Collisional effects destroy the phase memory in a short time compared to the excited state lifetime. The tip of the **R** vector relaxes toward the 3 axis a rate T_2^{-1} and then relaxes toward its equilibrium position $\mathbf{R}^e$ at the slower rate T_b^{-1}. (b) This situation applies when population decay is the only dephasing mechanism, and the tip of the **R** vector traces a curve in the 2-3 plane. (c, d) These situations occur when the driving fields and frame rotation frequency are off resonance. (c) For fast dephasing, **R** spirals in to the 3 axis before the population difference recovers. (d) When population decay dominates the dephasing process the tip of the **R** vector follows the whirlpool shaped trajectory. In the presence of relaxation, the magnitude of the Bloch vector is not generally conserved.

$$w(\infty) = \{-1\}\left(1 - \frac{\chi^2 T_2/T_b}{\Delta^2 + T_2^{-2} + \chi_0^2 T_2/T_b}\right). \quad (2.4.21c)$$

Again, in the T_1, T_2 approximation, we must replace the -1 in the braces in (2.4.21) with w^e and T_1 must replace T_b.

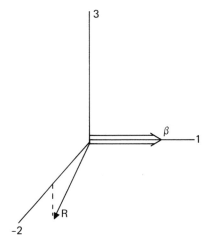

Fig. 2.5 The steady-state orientations of **β** and **R** in the presence of relaxation. The case illustrated is that of perfect resonance; otherwise, the components of **R** are given in (2.4.21). The steady-state situation evolves out of the transient cases in Figs. 2.2 and 2.3 in a manner reminiscent of Fig. 2.4.

Some effects induced by the external fields can be treated as relaxation phenomena. Perhaps the most important of these is photoionization, in which one state (usually the upper level $|b\rangle$) is coupled to a higher lying continuum by the incident fields as in Fig. 2.6. One result of this coupling is

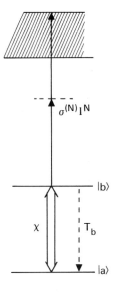

Fig. 2.6 A two-level system with a relaxation process induced by the external field. In this case, the wave causing the coupling between $|a\rangle$ and $|b\rangle$ described by the Rabi frequency χ also couples $|b\rangle$ to a continuum. The depopulation rate of $|b\rangle$ due to the external field is given by the product of a multiquantum cross section $\sigma^{(N)}$ and the Nth power of the light intensity.

2.4 The Master Equation and the Vector Model

an intensity-dependent component in the relaxation rate of that state. If the ionization process for state $|b\rangle$ can be described by (1.1.8), the relaxation rate of state $|b\rangle$ becomes

$$T_b^{-1} \to T_b^{-1} + \sigma^{(n)} I^n \qquad (2.4.22a)$$

and the transverse relaxation rate turns into

$$T_2^{-1} \to \tfrac{1}{2} T_b^{-1} + \tfrac{1}{2} \sigma^{(n)} I^n. \qquad (2.4.22b)$$

The instantaneous ionization rate becomes

$$\Gamma(t) = \sigma^{(n)} I^n \rho_{bb}(t). \qquad (2.4.22c)$$

Many resonant multiquantum ionization processes require more complete treatment. When both states $|a\rangle$ and $|b\rangle$ can be ionized with similar cross sections, interference and coherence phenomena must be incorporated explicitly in the theory.

The majority of the effects useful in nonlinear spectroscopy can be explained in terms of these solutions of the equation of motion for the Bloch–Feynman vector. Equations (2.4.13) can be used to transform out of the rotating reference frame, and to obtain the density matrix itself. Formula (2.1.8) then yields the expectation value of any operator, averaged over the ensemble. Shortcuts for calculating important quantities will appear in later sections.

When more than one driving frequency acts on a two-level system, the rotating reference frame cannot usually produce solutions in closed form. It is then usually necessary to expand the density matrix or the Bloch–Feynman vector as a Taylor series in the incident field amplitudes and a Fourier series in the incident frequencies, their sums, and differences. The Taylor–Fourier coefficients can be obtained from (2.4.3), (2.4.5), or (2.4.10) by the method of successive approximation. Since most nonlinear phenomena are at worst cubic in the coupling Hamiltonian, this process rapidly yields the desired result.

As an example of this technique, consider the case where two waves of frequency ω_1 and ω_2 interact with a two-level system with resonant frequency Ω. The interaction Hamiltonian is then

$$\tilde{\mathscr{H}}_1(t) = \hbar \begin{bmatrix} 0 & \chi_1 e^{-i\omega_1 t} \\ \chi_1 e^{i\omega_1 t} & 0 \end{bmatrix} + \hbar \begin{bmatrix} 0 & \chi_2 e^{-i\omega_2 t} \\ \chi_2 e^{i\omega_2 t} & 0 \end{bmatrix}, \qquad (2.4.23)$$

where

$$\chi_i = -\tilde{\mu}_{ab} \cdot \mathbf{E}_i e^{i\mathbf{k}_i \cdot \mathbf{r}}/\hbar,$$

and we shall expand the components of the density matrix as

$$\rho_{ab}(t) = [\rho_{ab}^{(1)}(\omega_1) + \rho_{ab}^{(3)}(\omega_1) + \cdots]e^{-i\omega_1 t} + [\rho_{ab}^{(1)}(\omega_2) + \rho_{ab}^{(3)}(\omega_2) + \cdots]e^{-i\omega_2 t}$$
$$+ \rho_{ab}^{(3)}(2\omega_1 - \omega_2)e^{-i(2\omega_1 - \omega_2)t} + \cdots, \quad (2.4.23a)$$

$$\rho_{bb} - \rho_{aa} = w = -1 + w^{(2)}(0) + w^{(2)}(\omega_1 - \omega_2)e^{-i(\omega_1 - \omega_2)t}$$
$$+ w^{(2)}(\omega_2 - \omega_1)e^{i(\omega_1 - \omega_2)t} + \cdots. \quad (2.4.23b)$$

The parentheses in (2.4.23) refer to the frequency at which that component is oscillating, while the superscript indicates its order in the applied amplitude. It is convenient to combine (2.4.5a) and (2.4.5b) and restate (2.4.5c) and (2.4.5d):

$$\dot{\rho}_{ab} + (i\Omega + T_2^{-1})\rho_{ab} = (i/2)w(\chi_1 e^{-i\omega_1 t} + \chi_2 e^{-i\omega_2 t}), \quad (2.4.24a)$$

$$\dot{w} + (w+1)/T_b = i\rho_{ba}(\chi_1 e^{-i\omega_1 t} + \chi_2 e^{-i\omega_2 t})$$
$$+ i\rho_{ab}(\chi_1^* e^{i\omega_1 t} + \chi_2^* e^{i\omega_2 t}), \quad (2.4.24b)$$

$$\rho_{ab} = \rho_{ba}^*. \quad (2.4.24c)$$

To obtain the first-order components, one inserts the zero-order component $\{-1\}$ from (2.4.23b) in (2.4.24a),

$$\rho_{ab}^{(1)}(\omega_1) = \frac{i/2\{-1\}\chi_1}{i(\Omega - \omega_1) + T_2^{-1}}, \quad (2.4.25a)$$

$$\rho_{ab}^{(1)}(\omega_2) = \frac{i/2\{-1\}\chi_2}{i(\Omega - \omega_2) + T_2^{-1}}, \quad (2.4.25b)$$

$$\rho_{ba}(-\omega_1) = \rho_{ab}^*(\omega_1), \quad (2.4.25c)$$

$$\rho_{ba}(-\omega_2) = \rho_{ab}^*(\omega_2). \quad (2.5.25d)$$

It is clear from (2.4.24b) that $w^{(1)} = 0$. The second-order components are calculated by substituting (2.4.25) in (2.4.24b),

$$w^{(2)}(0) = -T_b\{-1\}\left[\frac{T_2^{-1}\chi_1\chi_1^*}{(\Omega - \omega_1)^2 + T_2^{-1}} + \frac{T_2^{-1}\chi_2\chi_2^*}{(\Omega - \omega_2)^2 + T_2^{-2}}\right], \quad (2.4.26a)$$

$$w^{(2)}(\omega_1 - \omega_2) = -\frac{1}{2}\frac{\{-1\}\chi_1\chi_2^*}{-i(\omega_1 - \omega_2) + T_b^{-1}}\left[\frac{1}{i(\Omega - \omega_1) + T_2^{-1}}\right.$$
$$\left. + \frac{1}{-i(\Omega - \omega_2) + T_2^{-1}}\right], \quad (2.4.26b)$$

$$w^{(2)}(\omega_2 - \omega_1) = w^{(2)*}(\omega_1 - \omega_2), \quad (2.4.26c)$$

and again it is clear that the assumption $\rho_{ab}^{(2)} = 0$ is fulfilled. Equations (2.4.3) would give a nonzero value of $w^{(2)}(2\omega_i)$, but when $\omega_i T_b \gg 1$, $w^{(2)}(2\omega_i)$ is very

2.5 The Nonlinear Polarization Density and Nonlinear Susceptibility

small. In the present rotating-wave approximation, $w^{(2)}(2\omega_1) = 0$. Typical third-order components are obtained by substituting (2.4.26) in (2.4.24a),

$$\rho_{ab}^{(3)}(\omega_2)$$

$$= \frac{-i\{-1\}\chi_2/2}{(i\Omega - \omega_2) + T_2^{-1}} \left\{ \frac{T_b}{T_2} \left[\frac{|\chi_1|^2}{(\Omega - \omega_1)^2 + T_2^{-2}} + \frac{|\chi_2|^2}{(\Omega - \omega_2)^2 + T_2^{-2}} \right] \right.$$

$$\left. + \frac{|\chi_1|^2/2}{i(\omega_1 - \omega_2) + T_b^{-1}} \left[\frac{1}{T_2^{-1} - i(\Omega - \omega_1)} + \frac{1}{T_2^{-1} + i(\Omega - \omega_2)} \right] \right\}, \quad (2.4.27a)$$

$$\rho_{ab}^{(3)}(2\omega_1 - \omega_2) = \frac{-i\{-1\}\chi_1^2\chi_2/4}{[i(\Omega + \omega_2 - 2\omega_1) + T_2^{-1}][-i(\omega_1 - \omega_2) + T_b^{-1}]}$$

$$\times \left[\frac{1}{i(\Omega - \omega_1) + T_2^{-1}} + \frac{1}{-i(\Omega - \omega_2) + T_2^{-1}} \right]. \quad (2.4.27b)$$

Other frequency components can also be obtained; these two, however, are the ones of interest in the so called "moving grating" versions of four-wave mixing.

This density matrix or Bloch–Feynman vector formalism in exceedingly flexible. The matrix elements or vector components can be calculated in any desired basis and labeled with any number of variables. In particular, the expansions in (2.4.23) could have included the position of the ensemble as an argument, its wave vector, the resonant frequency, velocity, etc. The solutions in this section depend parametrically on all of these quantities.

2.5 THE NONLINEAR POLARIZATION DENSITY AND NONLINEAR SUSCEPTIBILITY

One response of the ensemble to the incident radiation takes the form of a dielectric polarization density **P** which acts as a source term in Maxwell's wave equation. The polarization density is related to the expectation value of the dipole moment operator and can be calculated from the density matrix by

$$\mathbf{P} = \mathcal{N}\langle \tilde{\boldsymbol{\mu}} \rangle = \mathcal{N} \text{Tr}(\tilde{\boldsymbol{\mu}}\rho), \quad (2.5.1)$$

where $\mathcal{N}$ is the number density in the ensemble. For the two-level systems of Section 2.4, the dielectric polarization density is

$$\mathbf{P}(t) = \mathcal{N}(\tilde{\boldsymbol{\mu}}_{ab}\rho_{ba} + \tilde{\boldsymbol{\mu}}_{ba}\rho_{ab}) \quad (2.5.2)$$

or

$$\mathbf{P}(t) = \mathcal{N} \text{Re}\{\tilde{\boldsymbol{\mu}}_{ab}(u + iv)e^{-i\omega t}\}, \quad (2.5.3)$$

where u and v are the transverse components of the Bloch–Feynman vector in the reference frame rotating at frequency ω.

In the steady state, this dielectric polarization density can be expanded as a power series in the incident amplitudes,

$$P_i = \sum_j \chi_{ij} E_j + \sum_{jk} \chi^{(2)}_{ijk} E_j E_k + \sum_{jkl} \chi^{(3)}_{ijkl} E_j E_k E_l + \cdots, \quad (2.5.4)$$

where the subscripts now refer to the Cartesian coordinates of the tensor quantity. It is convenient to expand **P** as a Fourier series in the incident frequencies, their sums, and differences, and to deal separately with the amplitudes of each frequency component. The coefficients of the nonlinear terms of this expansion are the nonlinear susceptibility tensor elements [7]. The total polarization density is

$$P_i(t) = \sum_\omega \tfrac{1}{2} P_i(\omega) e^{-i\omega t} + \mathrm{cc}, \quad (2.5.5a)$$

where, for example,

$$P_i(\omega_3) = \sum_j \chi_{ij}(-\omega_3, \omega_3) E_j(\omega_3)$$
$$+ \sum_{jkl} \chi^{(3)}_{ijkl}(-\omega_3, \omega_2, \omega_1, \omega_0) E_j(\omega_2) E_k(\omega_1) E_l(\omega_0) + \cdots. \quad (2.5.5b)$$

In systems with inversion symmetry, the even-order nonlinear susceptibilities must vanish. Thus, the lowest-order nonvanishing nonlinear susceptibility is third order—$\chi^{(3)}$.

The third-order nonlinear susceptibility is a fourth-rank tensor with three independent frequency arguments. The resulting polarization density is written in the now standard Maker–Terhune notation as

$$P_i^{(3)}(\omega_3) = \sum_{jkl} \chi^{(3)}_{ijkl}(-\omega_3, \omega_2, \omega_1, \omega_0) E_j(\omega_2) E_k(\omega_1) E_l(\omega_0), \quad (2.5.6)$$

where the summation is performed over all possible spatial components of the incident fields [2]. The four frequency arguments must sum to zero in this notation

$$\omega_0 + \omega_1 + \omega_2 - \omega_3 = 0, \quad (2.5.7)$$

and the frequency arguments and polarization indices of (2.5.6) can be freely permuted so long as their pairing is respected,

$$\chi^{(3)}_{ijkl}(-\omega_3, \omega_0, \omega_1, \omega_2) \equiv \chi^{(3)}_{iilkj}(-\omega_3, \omega_2, \omega_1, \omega_0)$$
$$\equiv \chi^{(3)}_{kijl}(\omega_1, -\omega_3, \omega_0, \omega_2). \quad (2.5.8)$$

As a result of these conventions, Eq. (2.5.6) becomes

$$P_i^{(3)}(\omega_3) = D \chi^{(3)}_{ijkl}(-\omega_3, \omega_2, \omega_1, \omega_0) E_j(\omega_2) E_k(\omega_1) E_l(\omega_0) + \mathrm{cc}, \quad (2.5.9)$$

2.6 Inhomogeneous Broadening

where D is the number of indistinguishable permutations of the argument—subscript pairs and the sum implied in the Einstein repeated index summation convention is taken over only *one* permutation.

The nonlinear susceptibility and dielectric polarization density resulting from the third-order density matrix term in (2.4.27b) is thus in Maker–Terhune notation

$$\chi^{(3)}_{1111}(\omega_2 - 2\omega_1, \omega_1, \omega_1, -\omega_2)$$

$$= \frac{\mathcal{N}\{1\}\mu^4/12}{[i(\Omega + \omega_2 - 2\omega_1) + T_2^{-1}][-i(\omega_1 - \omega_2) + T_b^{-1}]}$$

$$\times \left\{ \frac{1}{i(\Omega - \omega_1) + T_2^{-1}} + \frac{1}{i(\Omega - \omega_2) + T_2^{-1}} \right\}, \quad (2.5.10)$$

$$P_1^{(3)}(2\omega_1 - \omega_2) = 3\chi^{(3)}_{1111}E_1^2(\omega_1)E_1^*(\omega_2) = 3\chi^{(3)}_{1111}E_1^2(\omega_1)E_1(-\omega_2) \quad (2.5.11)$$

when all of the incident fields are polarized parallel.

A fourth-rank tensor can have as many as 81 independent elements. The symmetry of the medium, however, reduces this number to a more manageable proportion. The possible forms of $\chi^{(3)}$ for the 32 point groups are tabulated in Table 4.1; for an isotropic medium there are only four distinct nonzero tensor elements

$$\chi^{(3)}_{1111}, \quad \chi^{(3)}_{1122}, \quad \chi^{(3)}_{1212}, \quad \text{and} \quad \chi^{(3)}_{1221}$$

which must obey the condition

$$\chi^{(3)}_{1111} = \chi^{(3)}_{1122} + \chi^{(3)}_{1212} + \chi^{(3)}_{1221}. \quad (2.5.12)$$

2.6 INHOMOGENEOUS BROADENING

So far in this discussion, we have assumed that all of the atoms in the ensemble have the same resonant frequency. Such an ensemble is termed homogeneously broadened, and does occasionally occur in nature. More common, however, is the situation in which there is a distribution of resonant frequencies present in the total ensemble. Such an ensemble is termed inhomogeneously broadened as the width of the spectral lines generally reflects the width of the distribution of resonant frequencies rather than the width of the resonance of an individual atom.

The inhomogeneous broadening of spectral lines of a vapor results from the thermal distribution of axial velocities. A wave having frequency and wave vector ω and $\mathbf{k}$ in the laboratory frame will appear to have the frequency $\omega - \mathbf{k} \cdot \mathbf{v}$ in the rest frame of an atom moving with velocity $\mathbf{v}$.

Correspondingly, the resonant frequency of that atom would appear to be $\Omega + \mathbf{k} \cdot \mathbf{v}$ in the laboratory frame rather than Ω. In calculating the polarization density resulting from an interaction with an ensemble of moving atoms, the density matrix must be parametrized by the atomic position and velocity and the integral performed over the Maxwell–Boltzmann velocity distribution.

$$\mathbf{P} = \mathcal{N} \int \int \int_{-\infty}^{\infty} G(v_x, v_y, v_z) \boldsymbol{\mu}_{ba} \rho_{ab}(v_x, v_y, v_z) \, dv_x \, dv_y \, dv_z + \text{cc}, \quad (2.6.1)$$

where

$$G = \left(\frac{1}{\sqrt{\pi} v_0}\right)^3 \exp\left(-\frac{v_x^2 + v_y^2 + v_z^2}{v_0^2}\right) \quad (2.6.2)$$

represents the velocity distribution; $v_0 = \sqrt{2kT/m}$ is the RMS thermal velocity and $\iiint G(v_x, v_y, v_z) \, dv_x \, dv_y \, dv_z = 1$. The density matrix element for a particular velocity group $\rho_{ab}(v_x, v_y, v_z)$ can be readily calculated in the rest frame of the atom by the techniques of Section 2.4, using the substitution $\omega_i - \mathbf{k}_i \cdot \mathbf{v} \to \omega_i$ for the incident frequencies. Alternatively, the time derivatives in the master equations can be replaced by convective derivatives $d/dt = \partial/\partial t + \mathbf{v} \cdot \nabla$ and the problem solved in the laboratory frame. The integral in (2.6.1) can be difficult to evaluate, especially when there are several waves present propagating in different directions or when the Doppler width $\Omega_D = \Omega v_0/c$ is comparable with the homogeneous width T_2^{-1}.

In condensed phases, inhomogeneous broadening results from differences in the environments occupied by the resonant two-level systems. The distribution of resonant frequencies can be parametrized by a one-dimensional distribution function $g(\Omega, \Omega_0, \Omega_D)$, described by an average frequency Ω_0, a standard deviation Ω_D, and which obeys

$$\int_{-\infty}^{\infty} g(\Omega, \Omega_0, \Omega_D) \, d\Omega = 1. \quad (2.6.3)$$

The dielectric polarization is then

$$\mathbf{P} = \mathcal{N} \int_{-\infty}^{\infty} g(\Omega, \Omega_0, \Omega_D) \boldsymbol{\mu}_{ba} \rho_{ba}(\Omega) \, d\Omega + \text{cc}, \quad (2.6.4)$$

where the density matrix element has been labeled by the resonant frequency. This simpler treatment can also be applied to experiments in gases where only one laser beam is involved. For gases, the distribution function is a Gaussian with standard deviation $\Omega_D = \Omega_0 v/c$:

$$g(\Omega, \Omega_0, \Omega_D) = (\sqrt{\pi} \Omega_D)^{-1} \exp[-(\Omega - \Omega_0)^2 / \Omega_D^2]. \quad (2.6.5)$$

The integral in (2.6.4) is greatly simplified when $\Omega_D \gg T_2^{-1}$. When the element of the density matrix appearing in (2.6.4) contains only one factor

2.6 Inhomogeneous Broadening

of the form $[i(\Omega - \omega) + \Gamma]^{-1}$, the integral can be evaluated by noting

$$\text{Re}[i(\Omega - \omega) + \Gamma]^{-1} \xrightarrow[\Gamma \to 0]{} \pi\delta(\Omega - \omega), \qquad (2.6.6a)$$

$$\text{Im}[i(\Omega - \omega) + \Gamma]^{-1} \xrightarrow[\Gamma \to 0]{} -(\Omega - \omega)^{-1}. \qquad (2.6.6b)$$

Thus, the steady-state polarization responsible for the absorption of an inhomogeneously broadened line is obtained using (2.4.21b),

$$\text{Im}\,\rho_{ab}(\Omega) = \frac{\{-1\}\chi_0}{(1 + \chi^2 T_b T_2)^{1/2}} \frac{(T_2^{-1} + \chi_0^2 T_b/T_2)^{1/2}}{(\Omega - \omega)^2 + T_2^{-2} + \chi_0^2 T_b/T_2}$$

$$\xrightarrow[T_2 \to \infty]{} \frac{\pi\{-1\}\chi_0}{(1 + \chi_0^2 T_b T_2)^{1/2}} \delta(\Omega - \omega), \qquad (2.6.7)$$

and performing the integral in the rotating frame for $\chi_0 = \mu_{ab} E_0(\omega)/\hbar$,

$$\text{Im}\,\mathbf{P}(\omega) = \frac{\mathcal{N}\{-1\}\pi\mu_{ab}^2 E_0/\hbar}{(1 + \chi_0^2 T_b T_2)^{1/2}} \int_{-\infty}^{\infty} g(\Omega, \Omega_0, \Omega_D)\delta(\Omega - \omega)\,d\Omega$$

$$= \frac{\mathcal{N}\{-1\}\pi\mu_{ab}^2/\hbar}{(1 + \chi_0^2 T_b T_2)^{1/2}} g(\omega, \Omega_0, \Omega_D) E_0(\omega). \qquad (2.6.8)$$

As expected, the line shape reflects the distribution of resonant frequencies. The magnitude of the polarization density is linearly proportional to the incident field amplitude only when $\chi_0^2 T_b T_2 \ll 1$. As the interaction becomes stronger, the response of the medium saturates. Of course, when the interaction is so strong that $\chi_0^2 T_b \approx \Omega_D^2 T_2$, this approximation breaks down.

When the distribution function is so wide that $g(\Omega, \Omega_0, \Omega_D)$ can be factored out of the integral in (2.6.4), contour integration in the complex plane can be used to evaluate the nonlinear polarization for $0 < \Gamma < \Omega_D$. For example, the polarization responsible for near-degenerate four-wave mixing in an inhomogeneously broadened absorbing medium can be expressed using (2.4.27b) and (2.6.4). Factoring out the distribution function and applying the residue theorem with a contour closed in the lower half-plane, we obtain Yajima's Rayleigh resonance formula [8]:

$$P^{(3)}(2\omega_1 - \omega_2)$$

$$= \mathcal{N} g(\omega_1, \Omega_0, \Omega_D) \int_{-\infty}^{\infty} \mu_{ba} \rho_{ab}^{(3)}(\Omega)\,d\Omega$$

$$= \frac{\mathcal{N}\pi\{-1\}\mu_{ab}^4 g(\omega_1, \Omega_0, \Omega_D)/2}{[i(\omega_2 - \omega_1) + T_2^{-1}][i(\omega_2 - \omega_1) + T_b^{-1}]} E^2(\omega_1) E^*(\omega_2). \qquad (2.6.9)$$

These line shapes are illustrated schematically in Fig. 2.7.

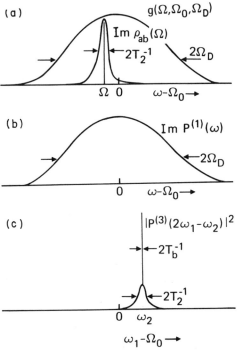

Fig. 2.7 The effect of inhomogeneous broadening upon the linear and nonlinear polarizations densities. (a) The distribution of resonant frequencies in the inhomogeneously broadened ensemble is described by $g(\Omega, \Omega_0, \Omega_D)$, while the absorption linewidth characteristic of a member with resonance frequency Ω is given by Im $\rho_{ab}(\Omega)$. (b) As a function of the laser frequency ω, the line shape for linear absorption reflects the distribution function. The magnitude of the nonlinear polarization density continues to reflect the characteristics of the members of ensemble. The polarization density for Rayleigh resonant four-wave mixing is given in (2.6.9) and the magnitude of the detected signal has the line shape shown in (c). The two components are centered where the two interacting laser frequencies are equal, independent of Ω_0. The stronger peak has width $2T_b^{-1}$, with wings extending to frequency $2T_2^{-1}$.

2.7 EFFECTIVE OPERATORS FOR MULTIQUANTUM TRANSITIONS

The two-level model and rotating-wave approximation apply rigorously only when the levels involved are coupled by a single quantum transition. In that case, the interaction Hamiltonian and dipole operator are

$$\mathcal{H}_1(t) = -\boldsymbol{\mu}_{ab} \cdot \mathbf{E}(t) \quad \text{and} \quad \tilde{\boldsymbol{\mu}} = \boldsymbol{\mu}_{ab}, \qquad (2.7.1)$$

where $\boldsymbol{\mu}_{ab} = \langle a|\tilde{\boldsymbol{\mu}}|b\rangle = \langle a|e\mathbf{r}|b\rangle$. Multiquantum transitions often require the presence of a third-level in the system. Heitler, however, pointed

2.7 Effective Operators for Multiquantum Transitions

out that when none of the incident frequencies is too near resonance with an additional level, operators can be defined that allow the two-level model to be extended into the multiquantum regime [9]. The most general algorithm for generating these operators is the projection operator formalism of Friedman and Wilson-Gordon [10].

The first stage in deriving effective two-level operators for multiquantum transitions is to define two projection operators, one of which specifies the two-level system being excited, and the other pointing to the remaining levels,

$$\tilde{P} = |a\rangle\langle a| + |b\rangle\langle b|, \qquad \tilde{Q} = \sum_{m \neq a,b} |m\rangle\langle m| \qquad (2.7.2)$$

and

$$\tilde{P} + \tilde{Q} = \tilde{I}. \qquad (2.7.3)$$

It is convenient to define the Fourier components of the incident field by

$$\mathbf{E}(t) = \tfrac{1}{2} \sum_{\omega_j} \{\mathbf{E}(\omega_j)e^{-i\omega_j t} + \mathbf{E}^*(\omega_j)e^{i\omega_j t}\} \qquad (2.7.4a)$$

and to include the very much smaller output amplitude as

$$\varepsilon(t) = \tfrac{1}{2}(\varepsilon_p e^{-i\omega_p t} + \varepsilon_p^* e^{i\omega_p t}). \qquad (2.7.4b)$$

The total matter-field interaction Hamiltonian then has the form

$$\tilde{V}_\varepsilon = \exp(i\mathcal{H}_0 t/\hbar)(-\tilde{\boldsymbol{\mu}} \cdot \mathbf{E}(t) - \tilde{\boldsymbol{\mu}} \cdot \varepsilon(t))\exp(-i\mathcal{H}_0 t/\hbar) \qquad (2.7.5)$$

in the interaction representation. Here $\mathcal{H}_0$ is the unperturbed Hamiltonian of the multilevel system with eigenstates $|a\rangle$, $|b\rangle$, and $|m\rangle$. If the adiabatic condition for the intermediate resonances

$$\frac{\partial |\mathbf{E}(\omega_j)|}{\partial t} \ll \mathbf{E}(\omega_j)|(\Omega_{ma} - \omega_1)| \qquad (2.7.6)$$

is fulfilled for all intermediate states $|m\rangle$ and all sums and differences of the incident frequencies $\omega_1 = \sum_j n_j \omega_j$, (i.e., $n_j = \pm 1, \pm 2, \ldots$), and if one particular sum or difference frequency is close to resonance

$$\Omega_{ba} - \sum_j p_j \omega_j \ll \Omega_{ma} - \omega_1, \qquad (2.7.7)$$

then the effective two-level interaction Hamiltonian is defined as

$$\tilde{\mathcal{H}}_1(t) \equiv \lim_{\varepsilon \to 0} \tilde{\mathcal{H}}_\varepsilon(t) \equiv \exp(-i\mathcal{H}_0 t/\hbar)\Big\{\tilde{P}\tilde{V}_\varepsilon\tilde{P} - (i/\hbar)\tilde{P}\tilde{V}_\varepsilon\tilde{Q}$$

$$\times \int^t \mathcal{T} \exp\left(-(i/\hbar)\int_{t'}^t \tilde{Q}\tilde{V}_\varepsilon\tilde{Q}\,dt''\right)\tilde{Q}\tilde{V}_\varepsilon\tilde{P}\,dt'\Big\}\exp(i\mathcal{H}_0 t/\hbar) \qquad (2.7.8)$$

in our original Schroedinger representation, where the chronologically ordered exponential in (2.7.8) is

$$\mathcal{T} \exp\left(-(i/\hbar)\int_{t'}^{t} \tilde{Q}\tilde{V}_{\varepsilon}\tilde{Q}\,dt''\right)$$
$$= 1 + (-i/\hbar)\int_{t'}^{t} \tilde{Q}\tilde{V}_{\varepsilon}(t'')\tilde{Q}\,dt''$$
$$+ (-i/\hbar)^2 \int_{t'}^{t}\int_{t'}^{t''} \tilde{Q}\tilde{V}_{\varepsilon}(t'')\tilde{Q}\tilde{V}_{\varepsilon}(t''')\tilde{Q}\,dt'''\,dt'' + \cdots \quad (2.7.9)$$

and setting $\epsilon = 0$ ensures that only the incident fields are effective in driving the ensemble. Equation (2.7.8) need only be evaluated to the lowest non-vanishing order in $\tilde{V}_{\varepsilon}(t)$.

The diagonal matrix elements of $\tilde{\mathcal{H}}_1$ correspond to optical Stark shifts of the levels $|a\rangle$ and $|b\rangle$. The resulting changes in the energy difference must be incorporated into the resonant frequency Ω,

$$\Omega \to (E_b - E_a + \mathcal{H}_{1bb}(t) - \mathcal{H}_{1aa}(t))/\hbar. \quad (2.7.10)$$

The off-diagonal elements induce transitions as before. In the rotating-wave approximation, only the frequency component of $\mathcal{H}_{1ab}^{(t)}$ nearest Ω need be retained.

The multiquantum dipole moment operator corresponding to the wave radiated at ω_p can be defined as a functional derivative

$$\tilde{\mu}(t) = \lim_{\varepsilon \to 0} \frac{\partial \tilde{\mathcal{H}}_{\varepsilon}(t)}{\partial \varepsilon(t)} = 2\lim_{\varepsilon \to 0}\left\{\frac{\partial \tilde{\mathcal{H}}_{\varepsilon}(t)}{\partial(\varepsilon_p e^{-i\omega_p t})} + \frac{\partial \tilde{\mathcal{H}}_{\varepsilon}(t)}{\partial(\varepsilon_p^* e^{i\omega_p t})}\right\}, \quad (2.7.11)$$

where only the off-diagonal matrix elements correspond to transitions between levels $|a\rangle$ and $|b\rangle$. Equation (2.7.11) need be evaluated only to the lowest order that gives a nonvanishing contribution at the necessary frequency.

As an illustration, the polarization responsible for coherent anti-Stokes Raman scattering can be calculated using this formalism. In CARS, the levels $|a\rangle$ and $|b\rangle$ have the same parity and are separated by a frequency nearly equal to the difference of the two incident frequencies ω_1 and ω_2 (see Fig. 1.9). The lowest-order nonvanishing interaction is bilinear in the incident amplitudes, and thus only the first term need be retained in the exponentials of (2.7.8) and (2.7.11). Substituting (2.7.5) into (2.7.8) and noting that $\langle a|\mu|b\rangle = 0$ and $\exp(-i\mathcal{H}_0 t/\hbar)|n\rangle = e^{-i\Omega_n t}|n\rangle$, we have for the off-

2.7 Effective Operators for Multiquantum Transitions

diagonal matrix element

$$\mathcal{H}_{I_{ba}}(t) = e^{-i\Omega_b t}\frac{-i}{4\hbar}e^{i\Omega_b t}\left\{\boldsymbol{\mu}_{bm}\cdot \mathbf{E}^*(\omega_2)e^{i\omega_2 t}\right.$$
$$\times e^{-\Omega_m t}\int_{-\infty}^{t}e^{i\Omega_m t'}\boldsymbol{\mu}_{ma}\cdot \mathbf{E}(\omega_1)e^{-i\omega_1 t'}e^{-i\Omega_a t'}\,dt' + \boldsymbol{\mu}_{bm}\cdot \mathbf{E}(\omega_1)e^{-i\omega_1 t}$$
$$\left.\times e^{-i\Omega_m t}\int_{-\infty}^{t}e^{i\Omega_m t'}\boldsymbol{\mu}_{ma}\cdot \mathbf{E}^*(\omega_2)e^{i\omega_2 t'}e^{i\Omega_a t'}\,dt'\right\}e^{i\Omega_a t}$$
$$= -\frac{1}{4\hbar}\left\{\frac{\boldsymbol{\mu}_{bm}\cdot \mathbf{E}^*(\omega_2)\boldsymbol{\mu}_{ma}\cdot \mathbf{E}(\omega_1)}{\Omega_{ma}-\omega_1} + \frac{\boldsymbol{\mu}_{bm}\cdot \mathbf{E}(\omega_1)\boldsymbol{\mu}_{ma}\cdot \mathbf{E}^*(\omega_2)}{\Omega_{ma}+\omega_2}\right\}e^{-i(\omega_1-\omega_2)t} \quad (2.7.12)$$

in the rotating-wave approximation to lowest nonvanishing order. This Hamiltonian can be expressed as

$$\mathcal{H}_{I_{ba}}(t) = -\tfrac{1}{4}\alpha^R_{\alpha\beta}(-\omega_1,\omega_2)E_\alpha(\omega_1)E^*_\beta(\omega_2)e^{-i(\omega_1-\omega_2)t} \quad (2.7.13)$$

in terms of the Raman susceptibility tensor employed by Maker and Terhune [2]

$$\alpha^R_{\alpha\beta}(-\omega_j,\omega_k) = \frac{1}{\hbar}\left\{\frac{\langle a|\tilde{\mu}_\alpha|m\rangle\langle m|\tilde{\mu}_\beta|b\rangle}{\Omega_{ma}-\omega_j} + \frac{\langle a|\tilde{\mu}_\beta|m\rangle\langle m|\tilde{\mu}_\alpha|b\rangle}{\Omega_{ma}+\omega_k}\right\}. \quad (2.7.14)$$

The most stringent adiabatic condition is

$$\partial E(\omega_j)/\partial t \ll E(\omega_j)(\Omega_{ma}-\omega_1). \quad (2.7.15)$$

As CARS is a steady-state effect, it is appropriate to employ the steady-state solutions (2.4.21) with

$$\chi_0 = \mathcal{H}_{I_{ba}}/\hbar \quad (2.7.16)$$

and

$$\chi_0^2 T_1 \ll T_2^{-1}$$

and the reference frame in this case rotates at frequency $\omega_1 - \omega_2$. The off-diagonal element of the dipole moment operator similarly becomes

$$\boldsymbol{\mu}_{ab} = \lim_{\varepsilon\to 0} 2\frac{\partial \tilde{\mathcal{H}}_\varepsilon(t)}{\partial(\varepsilon^* e^{i\omega_p t})} = \frac{1}{2\hbar}\left\{\frac{\boldsymbol{\mu}_{am}\cdot \mathbf{E}(\omega_1)\boldsymbol{\mu}_{mb}}{\Omega_{mb}+\omega_p} + \frac{\boldsymbol{\mu}_{am}\boldsymbol{\mu}_{mb}\cdot \mathbf{E}(\omega_1)}{\Omega_{mb}-\omega_1}\right\}e^{-i\omega_1 t}$$
$$= -\tfrac{1}{2}\alpha^R_{\alpha\beta}(-\omega_p,\omega_1)E_\beta(\omega_1)e^{-i\omega_1 t}, \quad (2.7.17)$$

where only the Fourier component at ω_1 contributes to the output wave at $\omega_p = 2\omega_1 - \omega_2$ and only the Fourier component of $\mathcal{H}_\varepsilon(t)$ at $\omega_1 - \omega_p$ is resonant. The relationship $|\Omega_{ba} - (\omega_1 - \omega_2)| \ll |\Omega_{ma} - \omega_1|$ has been employed to simplify the notation. Thus, substituting (2.7.13), (2.4.21), and (2.7.17) into (2.5.3), we obtain

$$P_\alpha^{(3)}(2\omega_1 - \omega_2)$$

$$= \frac{\mathcal{N}}{4\hbar} \frac{\alpha_{\alpha\beta}^R(-2\omega_1 + \omega_2, \omega_1)\alpha_{\gamma\delta}^R(-\omega_1, \omega_2) + \alpha_{\alpha\gamma}^R(-2\omega_1 + \omega_2, \omega_1)\alpha_{\beta\delta}^R(-\omega_1, \omega_2)}{\Omega - (\omega_1 - \omega_2) - i/T_2}$$

$$\times E_\beta(\omega_1)E_\gamma(\omega_1)E_\delta^*(\omega_2) \tag{2.7.18}$$

with

$$P_\alpha^{(3)}(t) = \tfrac{1}{2} P_\alpha^{(3)}(2\omega_1 - \omega_2) e^{-i(2\omega_1 - \omega_2)t} + \text{cc.} \tag{2.7.19}$$

The second term in the numerator results from the fact that either the β or γ components of $E(\omega_1)$ can appear in the Hamiltonian or dipole operator, and reflects the required permutation symmetry off resonance.

This result differs slightly from the exact solution for a four-level system in (2.8.9). If the $-\omega_1$ frequency component of μ_{ab} had been retained in (2.7.17) with $\omega_p = \omega_2$, the same calculation would have resulted in the polarization responsible for stimulated Raman gain. The polarization densities responsible for two-photon resonant third-harmonic generation and two photon resonant four-wave mixing are similar.

2.8 MULTIPLE RESONANCE EFFECTS

The formalism of Section 2.7 fails whenever more than two discrete energy levels are simultaneously near resonance. Five conditions of this sort appear in Fig. 2.8. Complications arise in these cases because of relaxation phenomena in the intermediate levels and because of correlations between different transitions. These complexities are best handled with the full density matrix formalism for a multilevel system. The definitions in Section 2.1 can readily be extended to the density matrices required to describe an n level system. Complete and explicit solutions have been obtained for the three-level case [11].

The mathematical complexity of this problem has stimulated the development of diagrammatic techniques. Because the familiar Feynman diagrams cannot correctly account for the relaxation phenomena that affect the wave function and its complex conjugate differently, special "double diagrams" have been developed by Gustafson and Borde. [12,13]. In these time-ordered diagrams, two separated vertical lines denote the propagators for the bra and ket parts of the density matrix defined in (2.1.6), or equivalently, for the left and right subscripts of ρ_{ij} with time increasing upward.

2.8 Multiple Resonance Effects

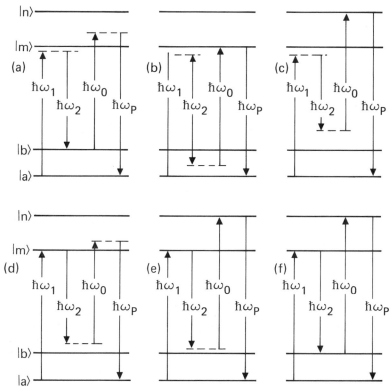

Fig. 2.8 Important multiple resonance processes in four-wave mixing. These diagrams represent only the magnitudes of level splittings and photon energies and should not be interpreted as describing population flows or time-ordered processes. (a) A typical Raman resonant four-wave mixing process describable by the effective operator technique. (b, c) Output frequency resonance conditions describable by different effective operators. (d) The input frequency drives a two-level resonance, and a third choice of projection operators must be made. (e, f) No effective two-level model can describe the double and triple resonance. A self-consistent description of four-wave mixing near these multiple resonance conditions requires the density matrix for a four-level system.

A typical Borde diagram appears in Fig. 2.9a. Two propagators are represented by the parallel vertical lines. The labels on line segments refer to energy eigenfunctions. The ρ_{uv} element of the density matrix would be represented by a pair of segments labeled u and v. Interactions with components of the applied fields oscillating as $e^{i\omega t}$ are represented by line segments sloping downward toward the vertex where they are attached to the propagators, while upward sloping lines denote interactions with components oscillating as $e^{-i\omega t}$. Segments to the left of the propagator correspond to

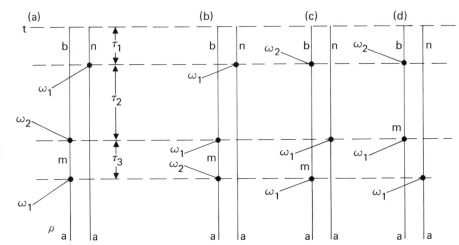

Fig. 2.9 Typical Borde diagrams for four-wave mixing. The two vertical propagator lines represent the ket and bra portions of the density matrix separately or equivalently the two subscripts of ρ_{uv}. Each of these diagrams begins with the system in the lower level. Interactions with frequency components of the field are represented by upward or downward sloping line segsegments in the manner described in the text. Where such segments connect with the propagators, the subscript is changed according to the dipole selection rules. These diagrams are time ordered, with time t at the top and the period between interactions given by τ_i. Each of these diagrams begins and ends with the same density matrix element, but the internal processes are distinct. The text describes the method for calculating the nonlinear polarization resulting from the process (a). The other diagrams also must be included in a description of the total third-order nonlinear polarization density as must many others.

waves propagating from left to right, while counterpropagating beams correspond to segments to the right of the propagators. Vertices occur on the left or right vertical line depending upon which of the two subscripts of ρ_{uv} is changed in the interaction. A new subscript labels the bar above the vertex. The top of the diagram corresponds to the time t, while the times of the prior interactions are $t - \tau_1$, $t - (\tau_1 + \tau_2)$, and $t - (\tau_2 + \tau_2 + \tau_3)$. To evaluate the contribution of a given diagram, integrals must be performed over the times τ_1, τ_2, and τ_3.

Between vertices, the elements of the density matrix propagator evolve as

$$\rho_{uv}^{(n)} = \rho_{uv}^{(n)}(\text{vertex}) e^{-i(\Omega_{uv} - i\Gamma_{uv})\tau_{n-m+1}}. \tag{2.8.1}$$

At a vertex of the right vertical line, the right subscript of the propagator is changed, the superscript indicating the order of the interaction (i.e., n) is increased by one and the propagator is multiplied by

$$\frac{+i\boldsymbol{\mu}_{vv'}}{2\hbar} \cdot \mathbf{E}(\omega_j) \exp\left[\pm i\omega_j\left(t - \sum_{m=1}^{N-n+1} \tau_m\right)\right], \tag{2.8.2}$$

2.8 Multiple Resonance Effects

where $\mu_{vv'}$ is the dipole matrix element between the eigenstates above and below the vertex. The minus sign is taken if the segment points up toward the vertex, the plus sign otherwise. The same rule applies to vertices on the left vertical line except that $-i\mu_{uu'}$ replaces $i\mu_{vv'}$ in (2.8.2) and the left subscript of ρ is altered. The complex amplitudes $\mathbf{E}(\omega_j)$ contain the position and phase of the wave and $\mathbf{E}(-\omega_j) = \mathbf{E}^*(\omega_j)$. The sum over m counts to $N - n + 1$ the number of vertices above the given vertex plus one, where N is the order of the nonlinearity. Thus, the lowest vertex of Fig. 2.9a corresponds to the transformation

$$\rho_{aa}^{(0)}(t-(\tau_1+\tau_2+\tau_3)) = \rho_{aa}^e \to \rho_{ma}^{(1)}(t-(\tau_1+\tau_2))$$

$$= \frac{i\mu_{ma}}{2\hbar} E(\omega_1) e^{-i\omega_1(t-[\tau_1+\tau_2+\tau_3])} e^{-i(\Omega_{ma}-i\Gamma_{ma})\tau_3} \rho_{aa}^e. \quad (2.8.3)$$

The propagator in Fig. 2.9a thus corresponds to the expression

$$\rho_{bn}^{(3)}(t,\tau_1,\tau_2,\tau_3) = \left(\frac{i\mu_{na} \cdot \mathbf{E}(\omega_1)}{2\hbar} e^{-i(\Omega_{bn}-i\Gamma_{bn})\tau_1} e^{-i\omega_1(t-\tau_1)} \right)$$

$$\times \left(\frac{i\mu_{bm} \cdot \mathbf{E}(-\omega_2)}{2\hbar} e^{-i(\Omega_{ba}-i\Gamma_{ba})\tau_2} e^{+i\omega_2(t-\tau_1-\tau_2)} \right)$$

$$\times \left(\frac{i\mu_{ma} \cdot \mathbf{E}(\omega_1)}{2\hbar} e^{-i(\Omega_{ma}-i\Gamma_{ma})\tau_3} e^{-i\omega_1(t-\tau_1-\tau_2-\tau_3)} \right) \rho_{aa}^e \quad (2.8.4)$$

and its contribution to the density matrix element is obtained by integrating over the possible histories indicated by the times $\tau_1, \tau_2,$ and τ_3. In the steady state, the limits of integration are 0 and ∞,

$$\rho_{bn}^{(3)}(t) = \int_0^\infty \int_0^\infty \int_0^\infty \rho_{bm}^{(3)}(t,\tau_1,\tau_2,\tau_3) \, d\tau_1 d\tau_2 d\tau_3$$

$$= \frac{\mu_{na} \cdot \mathbf{E}(\omega_1) \mu_{bm} \cdot \mathbf{E}(-\omega_2) \mu_{ma} \cdot \mathbf{E}(\omega_1) e^{-i(2\omega_1-\omega_2)t} \rho_{aa}^e}{8\hbar[\Omega_{bn}-(2\omega_1-\omega_2)-i\Gamma_{bn}][\Omega_{ba}-(\omega_1-\omega_2)-i\Gamma_{ba}][\Omega_{ma}-\omega_1-i\Gamma_{ma}]},$$
$$(2.8.5)$$

where the second equal sign applies only to the diagram of Fig. 2.9a.

To calculate the density matrix and nonlinear susceptibility corresponding to a given multiple resonant interaction, one must sum the contributions from all possible topological variations of the diagram in Fig. 2.9a. Three such diagrams appear in Figs. 2.9b–2.9d; other diagrams obtain with other initial conditions. The nonlinear source polarization must then be calculated using (2.5.1.). Three-hundred and eighty-four diagrams are required to evaluate the full third-order response of a four-level system. This diagrammatic technique can also be employed to identify processes leading to

Doppler-free resonances. The apparent laboratory frame phase shift of a moving atom $[\Omega_{uv} \pm (\omega_j \pm k_j v_z)]\tau_m$ must be substituted for $(\Omega_{uv} \pm \omega_j)\tau_m$ in the expression for ρ. The minus sign in front of k_j is taken for waves propagating left to right and denoted in the diagrams by line segments to the left of the vertical propagators. The plus sign is taken when the line segment extends to the right of the vertex. The integrand in 2.8.5 then has the form $\rho^{(N)}(t,\tau_1,\tau_2,\tau_3)e^{i\phi(v_z,\tau_1,\tau_2,\tau_3)}$, where $\phi(v_z,\tau_1,\tau_2,\tau_3) = v_z\sum_{m=1}^{N} k^{(m)}\tau_m$. If a combination of the τ_m's can be found for which $\sum k^{(m)}\tau_m = 0$, a Doppler-free resonance will remain after Eq. (2.8.5) has been averaged over v_z.

Lynch has provided a general formula for the third-order nonlinear susceptibility of a four-level system [14]. The master equations have the form

$$\dot{\rho}_{uv} + [i(\Omega_u - \Omega_v) + \Gamma_{uv}]\rho_{uv} = (-i/\hbar)[\tilde{\mathscr{H}}_1\rho]_{uv} + \Gamma_{ss}\delta_{uv}\delta_{vs}\rho_{ss}^e. \quad (2.8.6)$$

In (2.8.5) Γ_{uv} are the relaxation rates of the individual elements of the density matrix, with no assumptions made as to relationships among them. The diagonal terms of the density matrix relax toward their values at thermal equilibrium

$$\rho_{vv}^e = \exp(-\hbar\Omega_v/kT)\Big/\sum_v \exp(-\hbar\Omega_v/kT) \quad (2.8.7)$$

at the population decay rates. The off-diagonal elements relax toward zero at the transverse decay rates. In general,

$$\Gamma_{uv} = \tfrac{1}{2}(\Gamma_{uu} + \Gamma_{vv}) + \gamma_{uv}, \quad (2.8.8)$$

where γ_{uv} is the pure dephasing rate. More general assumptions concerning relaxation processes are possible, but the calculations then become prohibitively complicated.

The coupling Hamiltonian is $\tilde{\mathscr{H}}_1 = -\tilde{\boldsymbol{\mu}} \cdot \mathbf{E}(t)$, where the dipole matrix element vanishes between states of the same symmetry

$$\langle g|\tilde{\boldsymbol{\mu}}|r\rangle = \langle m|\tilde{\boldsymbol{\mu}}|n\rangle = 0$$

and $\mathbf{E}(t)$ appears in Eqs. (2.7.4a). The general result is so complex that special notation must be developed. We define first a nonlinear susceptibility tensor matrix element

$$\langle v|\tilde{\chi}^{(3)}_{\alpha\beta\gamma\delta}(-\omega_p,\omega_0,\omega_1,-\omega_2)|u\rangle$$

which corresponds to the nonlinear susceptibility provided solely by process beginning at state u and involving state v. The total third-order nonlinear susceptibility tensor can then be written as

$$\chi^{(3)}_{\alpha\beta\gamma\delta}(-\omega_p,\omega_0,\omega_1,-\omega_2) = \mathcal{N}\sum_{u,v}\langle v|\tilde{\chi}^{(3)}_{\alpha\beta\gamma\delta}(-\omega_p,\omega_0,\omega_1,-\omega_2)|u\rangle\rho_{uu}^e,$$

$$(2.8.9)$$

2.8 Multiple Resonance Effects

where ρ_{uu}^e is the density matrix of the system in thermal equilibrium, i.e., with the incident lasers turned off. The tensor matrix elements can be written as

$$\langle v|\tilde{\chi}_{\alpha\beta\gamma\delta}^{(3)}(-\omega_p,\omega_0,\omega_1,-\omega_2)|u\rangle = \frac{1}{48\hbar^3}(\tilde{\mu}_{un}\tilde{\mu}_{nv}\tilde{\mu}_{vm}\tilde{\mu}_{mu})$$

$$\times \left\{ \frac{1}{[\hat{\Omega}_{vu}-(\omega_1-\omega_2)](\hat{\Omega}_{nu}+\omega_2)}\left(\frac{\delta\gamma\beta\alpha}{\hat{\Omega}_{mu}-\omega_p}+\frac{\delta\gamma\alpha\beta[1+K_1(\Omega_3,\Omega_2)]}{\hat{\Omega}_{mu}^*+\omega_0}\right) \right.$$

$$+\frac{1}{[\hat{\Omega}_{vu}-(\omega_1-\omega_2)](\hat{\Omega}_{nu}-\omega_1)}\left(\frac{\gamma\delta\beta\alpha}{\hat{\Omega}_{mu}-\omega_p}+\frac{\gamma\delta\alpha\beta[1+K_1(\Omega_3,\Omega_1)]}{\hat{\Omega}_{mu}^*+\omega_0}\right)$$

$$+\frac{1}{[\hat{\Omega}_{vu}-(\omega_0-\omega_2)](\hat{\Omega}_{nu}+\omega_2)}\left(\frac{\delta\beta\gamma\alpha}{\hat{\Omega}_{mu}-\omega_p}+\frac{\delta\beta\alpha\gamma[1+K_1(\Omega_1,\Omega_3)]}{\hat{\Omega}_{mu}^*+\omega_1}\right)$$

$$+\frac{1}{[\hat{\Omega}_{vu}-(\omega_0-\omega_2)](\hat{\Omega}_{nu}-\omega_0)}\left(\frac{\beta\delta\gamma\alpha}{\hat{\Omega}_{mu}-\omega_p}+\frac{\beta\delta\alpha\gamma[1+K_1(\Omega_2,\Omega_1)]}{\hat{\Omega}_{mu}^*+\omega_1}\right)$$

$$+\frac{1}{[\hat{\Omega}_{vu}-(\omega_0+\omega_1)](\hat{\Omega}_{nu}-\omega_1)}\left(\frac{\gamma\beta\delta\alpha}{\hat{\Omega}_{mu}-\omega_p}+\frac{\gamma\beta\alpha\delta[1+K_1(\Omega_1,\Omega_3)]}{\hat{\Omega}_{mu}^*-\omega_2}\right)$$

$$+\frac{1}{[\hat{\Omega}_{vu}-(\omega_0+\omega_1)](\hat{\Omega}_{nu}-\omega_0)}\left(\frac{\beta\gamma\delta\alpha}{\hat{\Omega}_{mu}-\omega_p}+\frac{\beta\gamma\alpha\delta[1+K_1(\Omega_1,\Omega_2)]}{\hat{\Omega}_{mu}^*-\omega_2}\right)$$

$$+\frac{1}{[\hat{\Omega}_{vu}^*+(\omega_1-\omega_2)](\hat{\Omega}_{mu}^*-\omega_2)}\left(\frac{\alpha\beta\gamma\delta}{\hat{\Omega}_{nu}^*+\omega_p}+\frac{\beta\alpha\gamma\delta[1+K_2(\Omega_3,\Omega_2)]}{\hat{\Omega}_{nu}-\omega_0}\right)$$

$$+\frac{1}{[\hat{\Omega}_{vu}^*+(\omega_1-\omega_2)](\hat{\Omega}_{mu}^*+\omega_1)}\left(\frac{\alpha\beta\delta\gamma}{\hat{\Omega}_{nu}^*+\omega_p}+\frac{\beta\alpha\delta\gamma[1+K_2(\Omega_3,\Omega_1)]}{\hat{\Omega}_{nu}-\omega_0}\right)$$

$$+\frac{1}{[\hat{\Omega}_{vu}^*+(\omega_0-\omega_2)](\hat{\Omega}_{mu}^*-\omega_2)}\left(\frac{\alpha\gamma\beta\delta}{\hat{\Omega}_{nu}^*+\omega_p}+\frac{\gamma\alpha\beta\delta[1+K_2(\Omega_2,\Omega_3)]}{\hat{\Omega}_{nu}-\omega_1}\right)$$

$$+\frac{1}{[\hat{\Omega}_{vu}^*+(\omega_0-\omega_2)](\hat{\Omega}_{mu}^*+\omega_1)}\left(\frac{\alpha\gamma\delta\beta}{\hat{\Omega}_{nu}^*+\omega_p}+\frac{\gamma\alpha\delta\beta[1+K_2(\Omega_2,\Omega_1)]}{\hat{\Omega}_{nu}-\omega_1}\right)$$

$$+\frac{1}{[\hat{\Omega}_{vu}^*+(\omega_0+\omega_1)](\hat{\Omega}_{mu}^*+\omega_1)}\left(\frac{\alpha\delta\beta\gamma}{\hat{\Omega}_{nu}^*+\omega_p}+\frac{\delta\alpha\beta\gamma[1+K_2(\Omega_1,\Omega_3)]}{\hat{\Omega}_{nu}+\omega_2}\right)$$

$$\left.+\frac{1}{[\hat{\Omega}_{vu}^*+(\omega_0+\omega_1)](\hat{\Omega}_{mu}^*+\omega_0)}\left(\frac{\alpha\delta\gamma\beta}{\hat{\Omega}_{nu}^*+\omega_p}+\frac{\delta\alpha\gamma\beta[1+K_2(\Omega_1,\Omega_2)]}{\hat{\Omega}_{nu}+\omega_2}\right)\right\},$$

(2.8.10)

where the complex frequencies $\hat{\Omega}_{ij} = \Omega_i - \Omega_j - i\Gamma_{ij}$ have been introduced to conserve space, and the symbol $\alpha\beta\gamma\delta$ denotes the order of the Cartesian components of the product of the dipole matrix elements $\tilde{\mu}_{un}\tilde{\mu}_{nv}\tilde{\mu}_{vm}\tilde{\mu}_{mu}$. Two "correction factors" appear in this expression,

$$K_1(\Omega_j, \Omega_k) = \frac{i(\Gamma_{vn} - \Gamma_{vu} - \Gamma_{un}) + i(\Gamma_{mn} - \Gamma_{mu} - \Gamma_{un})(\hat{\Omega}_{vu} - \Omega_j)/(\hat{\Omega}_{mn} - \Omega_k)}{\hat{\Omega}_{vn} - \omega_p},$$

(2.8.11)

$$K_2(\Omega_j, \Omega_k) = \frac{i(\Gamma_{vn} - \Gamma_{vu} - \Gamma_{un}) + i(\Gamma_{mn} - \Gamma_{mu} - \Gamma_{un})(\hat{\Omega}_{vu}^* + \Omega_j)/\hat{\Omega}_{mn} - \Omega_k)}{\hat{\Omega}_{vm}^* - \omega_p},$$

(2.8.12)

with arguments $\Omega_1 = \omega_0 + \omega_1$, $\Omega_2 = \omega_0 - \omega_2$, and $\Omega_3 = \omega_1 - \omega_2$. In the damping approximation where pure dephasing is absent or the relaxation rates of the elements of the density matrix are the sums of the relaxation rates of the amplitudes of the corresponding wave functions, these "correction factors" vanish. The susceptibility in (2.8.9) accounts correctly for all of the nonlinear resonances—Raman, Rayleigh, two photon absorption, etc.—previously discussed and for the interferences between them. It does not account for inhomogeneous broadening for the saturation of the nonlinearity expected when effects beyond third order become significant. More importantly, (2.8.9) does properly account for the relaxation of subtle coherence effects encountered when the interacting levels are degenerate. Such phenomena have been reviewed by Omont and other authors and have become increasingly important as polarization and optical pumping techniques have grown in popularity [15].

2.9 THE WAVE EQUATION AND THE DETECTED SIGNAL

In calculating the nonlinear source polarization density $\mathbf{P}^{(3)}$, it has been convenient to suppress all notations of spatial dependence. The complex exciting field amplitudes in (2.3)–(2.8) depend parametrically upon the position

$$\mathbf{E}(\omega_j) = \mathbf{E}_0(r, \omega_j) e^{i\mathbf{k}_j \cdot \mathbf{r}}, \quad (2.9.1)$$

where $|\mathbf{k}_j| = n(\omega_j)\omega_j/c$ is the wave vector of the radiation interacting with the ensemble. Similarly, the polarization density defined in (2.5.1) to (2.5.5) also depends parametrically upon position

$$\mathbf{P}^{(n)}(\omega_p) = \mathbf{P}_0(r, \omega_p) e^{i\mathbf{k}_p \cdot \mathbf{r}}$$

and

$$\mathbf{P}(r, t) = \tfrac{1}{2} \sum_{\omega_p} \mathbf{P}_0(r, \omega_p) e^{i\mathbf{k}_p \mathbf{r}} e^{-i\omega_p t} + \text{cc}. \quad (2.9.2)$$

2.9 The Wave Equation and the Detected Signal

The wave vector $\mathbf{k}_p = \sum \mathbf{k}_j$, however, depends upon the wave vectors of the beams employed to excite the ensemble. This quantity can be determined by inserting (2.9.1) in the appropriate formula for $\mathbf{P}^{(n)}$. The form and magnitude of the output from a nonlinear spectroscopy experiment depends crucially upon the spatial distribution of the driving polarization.

The nonlinear output amplitude is obtained from Maxwell's wave equation

$$\nabla \times \nabla \times \mathbf{E}_s(\mathbf{r}, t) + \frac{n^2(\omega_s)}{c^2} \ddot{\mathbf{E}}_s(\mathbf{r}, t) = -\frac{4\pi}{c^2} \ddot{\mathbf{P}}^{(n)}(\mathbf{r}, t), \qquad (2.9.3)$$

when the appropriate form for $\mathbf{P}^{(n)}(\mathbf{r}, t)$ is inserted. Since the amplitude of the signal wave varies little within a wavelength of light, the wave equation (2.9.3) can be cast in terms of the slowly varying complex signal amplitude

$$ik_s \frac{\partial \mathbf{E}_s(r)}{\partial z} = -4\pi \frac{\omega_p^2}{c^2} \mathbf{P}_0^{(n)}(r, \omega_p) e^{-i(k_s - k_p)z}, \qquad (2.9.4)$$

where

$$\mathbf{E}_s(\mathbf{r}, t) = \tfrac{1}{2} \mathbf{E}_s(\mathbf{r}) e^{-i(\omega_p t - \mathbf{k}_s \cdot \mathbf{r})} + \text{cc} \qquad (2.9.5a)$$

and

$$|k_s| = n(\omega_p)\omega_p/c \qquad (2.9.5b)$$

and the z axis has been chosen parallel to $\mathbf{k}_p$.

By definition, there is no nonlinear spectroscopy signal amplitude entering the sample medium. Thus, the correct boundary condition on (2.9.4) is

$$\mathbf{E}_s(-\tfrac{1}{2}l, t) = \dot{\mathbf{E}}_s(-\tfrac{1}{2}l, t) = 0 \qquad (2.9.6)$$

for a sample filling the region $-\tfrac{1}{2}l < z < \tfrac{1}{2}l$. If the incident beams can be treated as uniform plane waves, and if the amplitude is much less than the incident amplitudes, Eq. (2.9.4) can be integrated directly to give the output signal amplitude

$$\mathbf{E}_s = -\frac{4\pi i}{n(\omega_p)} \frac{\omega_p}{c} l \mathbf{P}_0^{(n)}(\omega_p) \operatorname{sinc}((k_s - k_p)l/2) e^{-i(k_s - k_p)l/2}, \qquad (2.9.7)$$

where $\operatorname{sinc} x = \sin x/x$, l is the sample length. The dipole moments at various positions within the sample add coherently to produce an appreciable output wave only when the wave-vector matching condition

$$|\mathbf{k}_s - \mathbf{k}_p| \ll \pi/l \qquad (2.9.8)$$

is fulfilled [7]. For some types of nonlinear spectroscopy, this condition is always and automatically obeyed; in others, achieving a geometry that approximates $|\mathbf{k}_s - \mathbf{k}_p| = 0$ is the most significant technical problem.

The quantity physically detected in nonlinear laser spectroscopy is not an amplitude, but rather an optical intensity or a change in optical intensity. In techniques such as CARS and resonant harmonic generation, where the signal beam can be spatially or spectrally separated from other waves; the detected intensity is related to the signal amplitude and polarization by

$$I_S = \frac{n(\omega_p)c}{8\pi} |E_s|^2 = \frac{2\pi}{n(\omega_p)} \frac{\omega_p^2}{c^2} l^2 |P_0^{(n)}(\omega_p)|^2 \operatorname{sinc}^2[(k_s - k_p)l/2]. \quad (2.9.9)$$

A single isolated resonance as in (2.7.18) yields a symmetrical Lorentzian line shape $I_S \propto 1/[\{\Omega - (\omega_1 - \omega_2)\}^2 + 1/T_2^2]$. More commonly the contributions from different resonances and different processes interfere with one another optically to yield a somewhat more complicated spectrum proportional to the square of the absolute value of the total nonlinear susceptibility:

$$I_S \propto |\chi^{(n)}|^2.$$

In stimulated Raman spectroscopy and saturated absorption in optically thin media, both the transmitted probe laser and the nonlinear interaction contribute to the intensity at the detector

$$I(\omega_2) = [n(\omega_2)c/4\pi] |\mathbf{E}_s + \mathbf{E}_2|^2. \quad (2.9.10)$$

Since $E_S \ll E_2$ the detected change in the intensity due to the nonlinear interaction is

$$\Delta I = [n(\omega_2)c/4\pi] \operatorname{Re} \mathbf{E}_2 \cdot \mathbf{E}_s = -(\omega_2 l/c) \operatorname{Im} \mathbf{E}_2 \cdot \mathbf{P}_0^{(n)}(\omega_2) \quad (2.9.11)$$

and $\mathbf{E}_2$ can be assumed real.

Techniques which employ optical heterodyne detection inject a local oscillator wave with controlled magnitude, frequency, and phase into the detector along with the nonlinearly generated signal. The total intensity at the detector is then

$$I_T(t) = \frac{n(\omega_p)c}{4\pi} |\mathbf{E}_{LO}(t) + \mathbf{E}_S(t)|^2 = I_{LO}(t) + I_S + \frac{n(\omega_p)c}{4\pi} \operatorname{Re} \mathbf{E}_{LO}^*(t) \cdot \mathbf{E}_S(t), \quad (2.9.12)$$

where I_{LO} is the local oscillator intensity, and the signal intensity $I_S \ll I_{LO}$ can be obtained from (2.9.9). The heterodyne signal is

$$I_h(t) = [n(\omega_p)c/4\pi] \operatorname{Re}(\mathbf{E}_{LO}^*(t) \cdot \mathbf{E}_S(t))$$
$$= -(\omega_p l/c) \operatorname{Im}(\mathbf{E}_{LO}^* \cdot \mathbf{P}_0^{(n)}(\omega_p) e^{-i(\omega_p - \omega_{LO})t}), \quad (2.9.13)$$

where ω_{LO} and $\mathbf{E}_{LO}$ are the local oscillator frequency and complex amplitude, respectively. When the local oscillator and nonlinear signal have the same center frequency, an isolated Lorentzian line can give either a resonance or dispersion lineshape reflecting the imaginary or real part of the nonlinear susceptibility. Otherwise, the envelope of the electrical signal oscillating at $|\omega_p - \omega_{LO}|$ reflects $|P_0^{(n)}(\omega_p)|$ or $|\chi^{(n)}|$.

When the interacting beams cannot be treated as uniform plane waves, a more detailed analysis is necessary. Bjorklund has performed such an analysis for collinear TEM_{00} modes interacting by means of a third-order optical nonlinearity [16]. The most salient feature of this result is that tight focusing need not maximize the generated signal amplitude. Rather, the limiting signal amplitudes are approached whenever the beams are focused to a radius w_0 which fulfills

$$w_0^2 \lesssim 10l/k, \qquad (2.9.14)$$

where l is the sample length and k is a typical wave vector in the nonlinear medium. For more tightly focused beams, the increase in intensity at focus is compensated by a decrease in the interaction length. For many processes, tight focusing also tends to reduce signal levels by broadening the resonant linewidths.

Some nonlinear spectroscopy techniques do not detect a wave coherently radiated by the sample, but rather measure indirectly the energy deposited in the medium. The signals detected by these techniques are related to changes in the diagonal elements of the density matrix. For the two-level systems of Section 2.4, the detected quantity is proportional to

$$\int_{\text{pulse}} \int_{\text{sample}} (\hbar\Omega/2)(\rho_{bb} - \rho_{aa}) \, dV \, dt, \qquad (2.9.15)$$

where the spatial integrals are performed over the sample and the time integral is performed over the laser pulse.

2.10 THE RECIPE

The diverse phenomena of nonlinear spectroscopy can be accounted for using a relatively simple formalism. Since spectroscopy involves resonances, it is first necessary to identify the levels responsible. If there are more than two such levels, the results of Section 2.8 or the exact formulas of Brewer and Hahn are required [11]. If there are only two levels near resonance, the next stage is to identify the Hamiltonian operator that couples them to the external field and the dipole operator that allows the transition to radiate a detectable signal. These operators and the means for constructing them appear in Section 2.7.

If the coupling Hamiltonian is linear in the applied field, the master equations for the density matrix must be solved exactly or at least to third order in the Rabi frequency in Eq. (2.4.2). Generally, Hamiltonians that are nonlinear in the incident fields produce effects described by solutions that are linear in the Rabi frequency. The master equations and many useful solutions appear in Section 2.4.

Having solved the density matrix to the required order, the nonlinear polarization density can be calculated by the methods of Section 2.5 or the deposited energy or ionization evaluated directly. If necessary, an integral over the distribution of resonant frequencies encountered in an inhomogeneously broadened transition can be performed by the techniques of Section 2.6.

The spatial dependence of the nonlinear polarization must then be made explicit by the substitutions in (2.9.1) and (2.9.2), and the radiated signal amplitude calculated using (2.9.4). The signal amplitude is detected as a change in intensity at the detector as described by (2.9.11), (2.9.12), or (2.9.13).

The recipe can be simplified somewhat by incorporating prepackaged ingredients such as the third-order nonlinear susceptibility. Such simplifications will be attempted in later sections. The underlying physics, however, always is contained in the evolution of the density matrix and its coupling to the radiation field by means of Maxwell's wave equation. By introducing the underlying concepts at an early stage, we have hopefully minimized ambiguity.

REFERENCES

1. M. Sargent III, M. O. Scully, and W. E. Lamb, Jr., "Laser Physics," p. 79 ff. Addison-Wesley, Reading, Massachusetts, 1974.
2. P. D. Maker and R. W. Terhune, *Phys. Rev.* **137**, A801 (1965).
3. K. E. Jones and A. H. Zewail, Theory of optical dephasing in condensed phases, in "Advances in Laser Chemistry" (A. H. Zewail, ed.), pp. 196–222, Springer-Verlag, Berlin, 1978, and references therein. (See also Ref. 7, Chapter 1, this volume.)
4. W. E. Lamb, Jr., and T. M. Saunders, Jr., *Phys. Rev.* **119**, (1960).
5. C. Cohen-Tannoudji, Atoms in strong resonant fields, in "Frontiers in Laser Spectroscopy" (R. Balian, S. Haroche, and S. Liberman, eds.), pp. 3–104. North-Holland Publ., Amsterdam, 1977.
6. L. Allen and J. H. Eberly, "Optical Resonance and Two Level Atoms" pp. 28 ff and 52 ff. Wiley (Interscience), New York, 1975.
7. N. Bloembergen, "Nonlinear Optics." Benjamin, New York, 1965, and references therein.
8. T. Yajima, H. Souma, and Y. Ishida, *Phys. Rev. A* **17**, 324 (1978).
9. W. Heitler, "The Quantum Theory of Radiation," 3rd ed., Oxford Univ. Press, London and New York, 1954.
10. H. Friedman and A. D. Wilson-Gordon, *Opt. Commun.* **24**, 5 (1978); **26**, 151 (1978).
11. R. G. Brewer and E. L. Hahn, *Phys. Rev. A* **11**, 1641 (1975).

References

12. T. K. Yee and T. K. Gustafson, *Phys. Rev. A* **18**, 1597 (1978). Also T. K. Yee, T. K. Gustafson, S. A. J. Orvet, J. P. E. Taran, *Opt. Commun.* **23**, 1–7 (1977) and T. K. Yee and M. C. Gowar, *IEEE J. Quant. Electron.* **QE-18**, 437–441 (1982).
13. C. J. Borde, "Density Matrix Equations and Diagrams for High Resolution Spectroscopy" (to be published), S. A. J. Druet, J. P. E. Taran, and C. J. Borde, *J. Phys.* **40**, 819–840 (1979).
14. N. Bloembergen, H. Lotem, and R. T. Lynch, Jr., *Indian J. Pure Appl. Phys.* **16**, 151–158 (1978).
15. A. Omont, *Prog. Quant. Electron.* **5**, 69 (1977).
16. G. C. Bjoklund, *IEEE J. Quant. Electron.* **QE-11**, 287 (1975).

Chapter 3
SATURATION SPECTROSCOPY

Saturation was perhaps the first spectroscopically interesting nonlinear optical effect, "discovered" soon after the operation of the first gas laser by Javan in 1961. It was soon recognized by Bennet and by Lamb that the narrow resonances or "Lamb dips" that appeared at the center of inhomogeneously broadened gain lines interacting with counterpropagating laser beams resulted from "holes" burned in the Maxwell–Boltzmann velocity distribution. This phenomenon and its variations simultaneously provided a means for finding the center frequency of such a line and for removing the inhomogeneous linewidth [1].

Improved spectroscopic precision, better measurements of fundamental constants and of basic units, and new studies of collisional dynamics were only some of the exciting possibilities opened by saturated absorption and related techniques. Experimental and theoretical developments rapidly succeeded one another, making saturation spectroscopy now one of the most highly developed and best-understood nonlinear techniques. A number of books, review articles, and original sources elucidate the many phenomena expected and observed [1–8]. Our purpose here is not to summarize this extensive literature, but only to introduce the most salient features.

3.1 BURNING AND DETECTING HOLES IN A DOPPLER-BROADENED TWO-LEVEL SYSTEM

In the most common saturation spectroscopy geometry, two counter-propagating waves of the same frequency,

$$\mathbf{E}(t) = \tfrac{1}{2}\{\mathbf{E}_+ e^{-i(\omega t + kz)} + \mathbf{E}_- e^{-i(\omega t - kz)}\} + \text{cc}, \qquad (3.1.1)$$

3.1 Burning and Detecting Holes in a Doppler-Broadened Two-Level System

interact with an inhomogeneously broadened absorption line. If the pump wave (E_+) is much stronger than the probe amplitude E_-, the absorption and dispersion at the probe frequency can be calculated rather simply. In the steady state, the effect of the pump beam on the population difference is given by (2.4.21c). For molecules moving with axial velocity v in the positive z direction, the detuning from the pump frequency is

$$\Delta = \Omega - \omega + kv$$

and the Rabi frequency is

$$\chi_+ = -\boldsymbol{\mu}_{ab} \cdot \mathbf{E}_+/\hbar.$$

The weaker probe beam encounters an ensemble modified by the effect of the pump wave. The "hole" does not appear in the velocity distribution $g(v)$ itself, but rather in the difference of the velocity distributions for the upper and lower states

$$g_D(v) = -g(v)[\rho_{bb}(v) - \rho_{aa}(v)] = -g(v)w_+(v) \tag{3.1.2}$$

which is now not Maxwell–Boltzmann, but rather given by

$$g_D(v) = \{-1\} \frac{e^{-v^2/v_0^2}}{\sqrt{\pi} v_0} \left(1 - \frac{|\chi_+|^2 T_2/T_b}{(\Omega - \omega + kv)^2 + T_2^{-2} + |\chi_+|^2 T_b/T_2}\right) \tag{3.1.3}$$

(see Fig. 1.6). The off-diagonal elements of the density matrix that are responsible for the absorption and dispersion of the probe can be calculated from (2.4.21a) and (2.4.21b) with the probe detuning and Rabi frequency given by

$$\Delta = \Omega - \omega - kv, \qquad \chi_- = -\boldsymbol{\mu}_{ab} \cdot \mathbf{E}_-/\hbar$$

and $g_D(v)$ replacing $\{-1\}$ in the numerators,

$$u(v) - iv(v) = \frac{g_D(v)\chi_-}{(\Omega - \omega - kv) - iT_2^{-1}}. \tag{3.1.4}$$

In (3.1.4), the probe beam has been assumed weak enough that $\chi_-^2 T_b \ll T_2^{-1}$. The polarization is then calculated by integrating (3.1.4) over the axial velocity. Rather than present the general result in terms of the complex plasma dispersion function, we shall assume $T_2^{-1} \ll \Omega_D = v_0 \Omega/c$, factor the Gaussian out of the integral, and perform the integral over the Lorentzians using the residue theorem:

$$\mathbf{P} = \left\{ \int_{-\infty}^{\infty} \frac{\mathcal{N}\{-1\}e^{-v^2/v_0^2}/\sqrt{\pi}v_0}{\Omega - \omega - kv} dv + i\{-1\} \frac{\mathcal{N} e^{-(\Omega-\omega)^2/\Omega_D^2}}{\sqrt{\pi}\Omega_D} \right\} \frac{\mu_{ab}^2 \mathbf{E}_-}{\hbar} e^{-i(\omega t - kz)}$$

$$+ i\{-1\}|\chi_+|^2 \frac{e^{-(\Omega-\omega)^2/\Omega_D^2}}{2\Omega_D \hbar} \frac{\mathcal{N}\sqrt{\pi T_b} \mu_{ab}^2 \mathbf{E}_- e^{-i(\omega t - kz)}}{\Omega - \omega - (i/2)T_2^{-1}[1 + (1 + |\chi_+|^2 T_b T_2)^{1/2}]}.$$

$$\tag{3.1.5}$$

The term on the right proportional to $|\chi_+|^2$ contains the resonance width $T_2^{-1}[1 + (1 + |\chi_+|^2 T_b T_2)^{1/2}]$—the Lamb dip. Physically, this feature results from the fact that when $\omega \approx \Omega$, the probe beam interacts with atoms at $v = 0$, some of which have already been promoted into the upper state by the pump beam. The factor $\mathcal{N}$ in (3.1.5) is the number density of the ensemble.

Equation (2.9.7) gives the signal amplitude under the assumption that the medium is optically thin:

$$\mathbf{E}_s = \{-1\} \frac{2\pi^{3/2} i\omega l}{n(\omega)\Omega_D c}$$

$$\times \frac{\mathcal{N}|\chi_+|^2 T_b e^{-(\Omega-\omega)^2/\Omega_D^2} \mu_{ab}^2/\hbar}{\Omega - \omega - (i/2)T_2^{-1}[1 + (1 + |\chi_+|^2 T_b T_2)^{1/2}]} \mathbf{E}_- e^{-i(\omega t - kz)}. \quad (3.1.6)$$

Note that $\mathbf{k}_s - \mathbf{k}_p \equiv 0$. The transmitted amplitude at the detector is similarly obtained:

$$\mathbf{E}_2 = \{1 - \kappa_0 l\} \mathbf{E}_- e^{-i(\omega t - kz)},$$

where

$$\kappa_0 = \frac{4\sqrt{\pi}\mathcal{N}\mu_{ab}^2/\hbar}{n(\omega)c\Omega_D} e^{-(\Omega-\omega)^2/\Omega_D^2} \quad (3.1.7)$$

is the linear amplitude attenuation constant. Thus, the change in detected intensity according to (2.9.12) is

$$\frac{\Delta I}{I} = 2\left|\frac{E_s}{E_2}\right| = \{1 - \kappa_0 l\}^{-1}(-\{-1\}) \frac{2\pi^{3/2}\omega l \mu_{ab}^2/\hbar}{n(\omega)\Omega_D c} e^{-(\Omega-\omega)^2/\Omega_D^2}$$

$$\times \frac{\mathcal{N}|\chi_+|^2 T_2^{-1} T_b[1 + (1 + |\chi_+|^2 T_b T_2)^{1/2}]}{(\Omega - \omega)^2 + \tfrac{1}{4}T_2^{-2}[1 + (1 + |\chi_+|^2 T_b T_2)^{1/2}]^2}, \quad (3.1.8)$$

where the pump intensity is $I_+ = [n(\omega)c\hbar^2/4\pi\mu_{ab}^2]|\chi_+|^2$ and the probe intensity I_- has been substituted for $[n(\omega)c/4\pi]|E_-|^2$. The change in the Beer's law absorption constant is

$$\Delta\kappa = \kappa'(\omega,\omega)I_+$$

$$= \{-1\} \frac{4\pi^{3/2}\mathcal{N}\omega\mu_{ab}^2/\hbar}{n(\omega)\Omega_D c} e^{-(\Omega-\omega)^2/\Omega_D^2}$$

$$\times \frac{|\chi_+|^2 T_2^{-1} T_b[1 + (1 + |\chi_+|^2 T_b T_2)^{1/2}]}{(\Omega - \omega)^2 + \tfrac{1}{4}T_2^{-2}[1 + (1 + |\chi_+|^2 T_b T_2)^{1/2}]^2}. \quad (3.1.9)$$

The Kramers–Kronig relationships require this narrow resonance to induce a dispersion in the index of refraction that can be detected by providing a local oscillator wave in quadrature to the transmitted probe wave. Taking

3.1 Burning and Detecting Holes in a Doppler-Broadened Two-Level System

the real part of (3.1.6), the change in the refractive index due to saturated dispersion is

$$\Delta n = \{-1\} \frac{2\pi^{3/2} \mathcal{N} \mu_{ab}^2/\hbar}{n^2(\omega)\Omega_D} |\chi_+|^2 T_b e^{-(\Omega-\omega)^2/\Omega_D^2}$$

$$\times \frac{\Omega-\omega}{(\Omega-\omega)^2 + \tfrac{1}{4}T_2^{-2}[1+(1+|\chi_+|^2 T_b T_2)^{1/2}]^2}. \quad (3.1.10)$$

Both effects are contained in the third-order nonlinear susceptibility, which for parallel polarizations becomes

$$\chi^{(3)}_{1111}(-\omega,\omega,-\omega,\omega) = \frac{n^2(\omega)}{24\hbar^3}|\mu_{ab}|^4 \frac{\mathcal{N}\{-1\}}{\sqrt{\pi}\Omega_D} \frac{T_b e^{-(\Omega-\omega)^2/\Omega_D^2}}{\Omega-\omega-iT_2^{-1}}, \quad (3.1.11)$$

where higher order effects have been eliminated by dropping the intensity dependence of the denominators [3, 6]. These higher-order effects principally cause a power broadening of saturation spectroscopy resonances. To avoid such broadening and possible intensity dependent shifts, it is now conventional to extrapolate the widths to zero pump and probe intensity.

The quantity

$$I_{\text{sat}} = [n(\omega)c/4\pi][(\mu_{ab}^2/\hbar^2)T_2 T_b]^{-1} \quad (3.1.12)$$

which appears in various guises in (3.1.3) to (3.1.11) is the saturation intensity. A beam with power per unit area I_{sat} reduces the absorption of its resonant velocity group by a factor of 2. Reasonable spectroscopists employ intensities much less than I_{sat}. Additional decay channels increase the longitudinal relaxation rates and modify I_{sat}. When the lifetime of the upper and lower states are T_b and T_a, respectively, and when the rate of direct relaxation from $|b\rangle$ to $|a\rangle$ is A_{ab}, I_{sat} becomes

$$I_{\text{sat}} = [n(\omega)c/4\pi][(\mu_{ab}^2/\hbar^2)T_2(T_a + T_b - A_{ab}T_a T_b)]^{-1}. \quad (3.1.13)$$

At higher intensities, the size and width of the nonlinear resonance grows less rapidly than implied by (3.1.9). When the hole-burning effect is still dominant and the pump and probe beams have the same intensities, a number of authors have derived the formula

$$\frac{\kappa}{\kappa_0} = \frac{1}{a_+ + a_-}\left[1 + \left(\frac{(\Omega-\omega)^2 T_2^2 + 1}{1+(\Omega-\omega)^2 T_2^2 + 2\chi^2 T_b T_2}\right)^{1/2}\right] \quad (3.1.14)$$

for the linear absorption constant [1,2]. In (3.1.4), κ_0 is the low-intensity absorption constant from (3.1.7) and

$$a_\pm = [1 + |\chi|^2 T_b T_2 - (\Omega-\omega)^2 T_2^2$$
$$\pm \{|\chi|^4 T_b^2 T_2^2 - 4(\Omega-\omega)^2 T_2^2(1+|\chi|^2 T_b T_2)\}^{1/2}]^{1/2}$$

Additional complexities result from the inevitable spatial nonuniformities of the interacting waves. The fact that the transverse dimensions of the beams are not infinite introduces a transit time broadening that will be discussed in Section 3.3. Standing-wave effects result from the fact that the counterpropagating beams produce intensity maxima and minima along the z axis. Near an intensity minimum, an atom with low axial velocity can avoid interacting with the incident waves altogether, while near a maximum, the atoms interact more strongly than our spatially averaged hole-burning picture would indicate. The resulting effects are closely related to the grating dip phenomena and time-reversed wavefront generation discussed in Section 3.9. Feldman and Feld have pointed out that these effects can greatly modify the shape of the hole in the velocity distribution, but for a Doppler-broadened transition, the lowest-order effect on the probe absorption scales as the square of the pump intensity [8].

At very high pump intensities, the absorption and emission line-shape functions of the individual atoms are altered by the optical frequency fields. The resulting effects can be calculated by evaluating to first order the probe-field-induced dipole moment by means of master equations in which the pump field is correctly treated to all orders. The dramatic changes in the line-shape function for a single velocity group is illustrated in Fig. 3.1. In the strong field limit, a homogeneously broadened absorption line can actually amplify the probe wave [9].

The reason for these dramatic changes in the line shape can be seen from the inset in Fig. 3.1b. The strong probe wave shifts and splits the energy eigenstates $|a\rangle$ and $|b\rangle$ by means of the Autler–Townes effect [10]. The new energy eigenstates of the coupled atom–pump wave system become

$$E_a^\pm = E_a + \tfrac{1}{2}\Delta \pm [(\tfrac{1}{2}\Delta)^2 + |\tfrac{1}{2}\chi|^2]^{1/2}, \qquad (3.1.15a)$$

$$E_b^\pm = E_b - \tfrac{1}{2}\Delta \pm [(\tfrac{1}{2}\Delta)^2 + |\tfrac{1}{2}\chi|^2]^{1/2}, \qquad (3.1.15b)$$

and thus, the four transitions with the three frequencies indicated in the inset are possible [1, 2, 11],

$$\Omega_\pm = \Omega + \Delta \pm [\Delta^2 + |\chi|^2]^{1/2}, \qquad (3.1.16a)$$

$$\Omega_0 = \Omega + \Delta, \qquad (3.1.16b)$$

where $\Omega = (E_b - E_a)/\hbar$ and $\Delta = \omega_+ - (E_b - E_a)/\hbar$ is the pump detuning. The frequency of maximum probe absorption is Ω_- while the amplification is possible at $2\omega_+ - \Omega_-$ due to a three-photon process involving the absorption of two pump photons at ω_+ and amplification of the probe at $2\omega_+ - \Omega_-$.

In the Doppler-broadened system in which the pump and probe detunings are velocity dependant, $\Delta_\pm = \omega - \Omega \pm kv$, some of the dramatic effects

3.1 Burning and Detecting Holes in a Doppler-Broadened Two-Level System

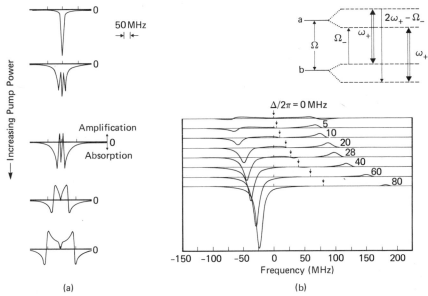

Fig. 3.1 Probe wave absorption and amplification for a homogeneously broadened two-level system interacting with a pump wave. (a) The pump wave is on-resonance. At top is the line shape obtained with zero pump intensity. The second through fifth traces show the effects of pump intensities of 26, 47, 130, and 560 mW/cm² on the sodium 3S(F = 2) → 3P$_{3/2}$(F = 3) transition in an atomic beam. The line shape ceases to be representable as a sum of Lorentzians and perturbation techniques break down. (b) The pump field is constant at 560 mW/cm² ($\chi_+/2\pi = 66$ MHz) and the pump detuning is indicated for each trace. The inset shows the possible transitions among the Autler–Townes split levels including the one at $2\omega_+ - \Omega_-$ which is responsible for the amplification of the probe. In saturation spectroscopy, these splittings and the consequent change of the absorption line shape for a single velocity group produce "coherent effects" which alter the observed line shape (from Ref. [9]).

average out. The only actual resonance condition is $\omega = \Omega$—exactly the Lamb dip condition; but each of the transitions in (3.1.16) contributes to the signal. Baklanov and Chebotayev, as well as Haroche and Hartman, have obtained equivalent formulas for the alteration in the line shape of the main Lamb dip due to these coherence effects [11–13]. The absorption constant for the weak probe beam becomes under the same assumptions that led to (3.1.9),

$$\frac{\kappa}{\kappa_0} = 1 + \frac{\Delta\kappa}{\kappa_0} + |\chi_+|^2 [1 - (1 + |\chi_+|^2 T_w T_2)^{-1/2}]$$

$$\times \mathrm{Re}\left\{ -\frac{f(-\Delta - iT_2^{-1}(1 + |\chi_+|^2 T_2 T_w)^{1/2})}{-2\Delta + iT_2^{-1}(1 + (1 + |\chi_+|^2 T_2 T_w)^{1/2})} \right\}, \quad (3.1.17)$$

where

$$f(x) =$$

$$\frac{(3x + \Delta + iT_2^{-1})(x + iT_2^{-1})(2x + iT_2^{-1})}{(3x + \Delta + iT_2^{-1})(x - \Delta + iT_2^{-1})(2x + iT_a^{-1})(2x + iT_b^{-1}) + (2x + iT_2^{-1})^2|\chi_+|^2}$$
(3.1.18)

$$T_w = T_a + T_b - A_{ab}T_aT_b, \quad \Delta = \omega - \Omega,$$ (3.1.19)

and $\Delta\kappa$ is obtained from (3.1.9) [2]. The third term in (3.1.17) results entirely from these "coherence effects"; its coefficient scales as $(\chi_+)^4$ at low pump intensity. Thus, the contribution of coherence effects can be minimized by using the least possible laser intensity. The importance of the coherence effects is a maximum when the two interacting levels have the same lifetime and pure dephasing is absent. When the lifetimes differ greatly, or pure dephasing or laser-phase fluctuations dominate the linewidth, coherence effects are negligible. The alterations in the Lamb dip line shape and magnitude due to these effects are illustrated in Fig. 3.2. Qualitatively, the main result of adding these coherence terms is that the probe wave absorption

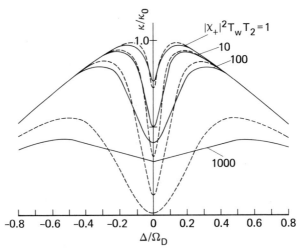

Fig. 3.2 The shape of an inhomogeneously broadened absorption line as indicated by a weak probe beam in the presence of a strong counterpropagating pump wave. The dip at $\Delta = 0$ is the Lamb dip. The dotted lines indicate the profiles predicted by our simple hole-burning model while the solid lines include the "coherence effects." The numbers on each trace indicate the degree of pump saturation; and T_w represents the relaxation time of the population difference which can be T_b, T_1, or $T_a + T_b - A_{ab}T_aT_b$ depending on the complexity of the decay process. Reasonable experiments are usually performed with pump saturation parameter values less than unity (from Ref. [2]).

3.2 CROSSOVER RESONANCES AND POLARIZATION SPECTROSCOPY

can never go to zero no matter how strong the pump wave. In most reasonable spectroscopy experiments, the coherence effects alter the depth and width of the Lamb dip by less than 20%.

The ideal two-level system is not often encountered in spectroscopy; normally, additional levels—sometimes degenerate—lie within the Doppler width of the prime transition. These additional levels often produce effects that cannot be explained within a two-level model of saturated absorption. In the spirit of this monograph, we shall deal here only with the basic features of the commoner and more useful effects, referring to the literature for the more esoteric cases [1–8].

The most common effect that occurs when two transitions having a common level interact with counterpropagating beams with the same frequency is illustrated in Fig. 3.3. The pump beam saturates the $|a\rangle - |b\rangle$ and $|a\rangle - |c\rangle$ transitions creating anomalous peaks or holes in the velocity

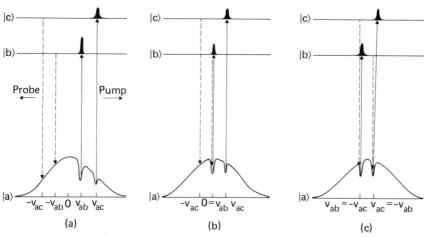

Fig. 3.3 Bennet holes, the Lamb dip, and crossover resonance for two coupled transitions. The pump wave is indicated by solid lines while the probe waves is represented by dashed lines. There is a Bennet hole in the lower state velocity distribution for each transition with center velocity given by (3.2.1), while the probe beam absorption depends upon the population difference at the negative of the hole velocity. (a) The waves are off-resonance and the probe beam absorption is unperturbed. (b) The Lamb dip condition for the $|a\rangle - |b\rangle$ transition. The pump and probe beams interact with the same velocity group. (c) The velocity group pumped by one transition is probed on the *other* transition, the result is a crossover peak at the frequency given in (3.2.2).

distribution of the common level. If the pump beam is propagating in the positive z direction, the anomalies in the velocity distribution are centered at

$$v_{ab}/c = (\omega - \Omega_{ab})/\omega \quad \text{and} \quad v_{ac}/c = (\omega - \Omega_{ac})/\omega. \tag{3.2.1}$$

The counterpropagating probe beam interacts with velocity groups at $-v_{ab}$ and $-v_{ac}$. When $\omega = \Omega_{ab}$ or $\omega = \Omega_{ac}$ the pump and probe interact with the same $v = 0$ velocity group and a normal inverted Lamb dip results. When

$$v_{ab} = -v_{ac},$$

however, the counterpropagating probe beam absorption due to one transition is altered as a result of the anomaly produced by the effect of the pump wave on the *other* transition which gives rise to a *crossover* dip in the absorption exactly halfway between the Lamb dips corresponding to $|a\rangle - |b\rangle$ and $|a\rangle - |c\rangle$ transitions:

$$\omega_x = \tfrac{1}{2}(\Omega_{ab} + \Omega_{ac}). \tag{3.2.2}$$

The width of the resonance is roughly equal to the average of the widths of the Lamb dips of the primary transitions, and if no other effect occurs, the size of the dip is proportional to the geometric mean of the Lamb dip amplitudes [3, 4].

There is, however, an additional effect that occurs when the upper level is in common as in Fig. 3.4a. Atoms excited into the common level can decay either into state $|a\rangle$ or into state $|c\rangle$. Atoms returning to their initial state can again be pumped into the common level by the strong beam, but atoms reaching the *other* state can be removed only by the weak probe beam and by slow relaxation processes. Collisions are often weak enough that the atoms reaching unpumped velocity regions retain the initial velocities that brought them into resonances with the pump wave. The result is an anomalous *increase* in the population of the velocity groups corresponding to the Bennet holes in the other level. A probe wave that interacts with these velocity groups encounters absorption that is increased as a result of the population anomaly in the lower level, but decreased as a result of the anomaly in level $|b\rangle$. Depending upon the relaxation rates, the crossover peak can be either positive (i.e., less absorption) or negative (increased absorption). Since highly excited states generally relax more rapidly than lower levels, the inverted case is more common [3].

One must also note that the optical pumping process described in the preceding paragraph decreases the effective saturation intensity I_{sat} making the true Lamb dips stronger, but complicating and widening their profiles. The extra complication results from the fact that a hole burned in the lower state velocity distribution may relax so slowly that velocity changing colli-

3.2 Crossover Resonances and Polarization Spectroscopy

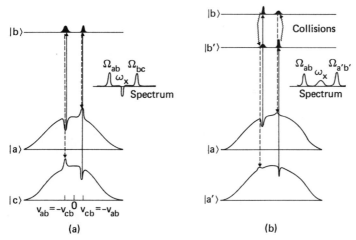

Fig. 3.4 Crossover resonances due to optical pumping and collisions. (a) Two low-lying states are thermally populated. The pump wave depopulates one velocity group in each state, but atoms decaying from the upper level produce peaks in the velocity distribution at the velocity corresponding to the hole in the *other* lower state. When the probe wave is resonant with a velocity group with an increased population, the absorption is *increased* and an inverted crossover peak results. (b) Selection rules prohibit direct decays or absorptions from state $|b\rangle$ to $|a'\rangle$ or $|b'\rangle$ to $|a\rangle$. Collisions, however, can transfer population from $|b'\rangle$ to $|b\rangle$ and vice versa without randomizing velocity. The resulting crossover resonances can be positive or negative.

sions have time to broaden it and shift its center. No complete theory of such spectral diffusion phenomena has yet been presented.

Crossover peaks can even occur when the selection rules imply that two transitions have no level in common. Such a case is diagrammed in Fig. 3.4b. The optically allowed transitions are $|a\rangle - |b\rangle$ and $|a'\rangle - |b'\rangle$. Collisions, spontaneous emission, or other perturbations can, however, mix levels $|b\rangle$ and $|b'\rangle$ without disturbing the atomic velocities. The result is that the peak in the velocity distribution for level $|b\rangle$ induced by the pump wave creates a corresponding anomaly in the population of level $|b'\rangle$ at the same velocity: $v = c(\omega - \Omega_{ab})/\omega$. A probe wave interacting with $|a'\rangle - |b'\rangle$ transition then senses a perturbed velocity distribution when the pump and probe laser frequency is $\omega_x = (\Omega_{a'b'} + \Omega_{ab})/2$. The resulting crossover peak can be either positive or negative, but the more common case is that of the negative resonance that results from a population peak in level $|a'\rangle$ resulting from the spontaneous decay of atoms in level $|b'\rangle$, etc. Crossover peaks can have other bizarre properties, some of which are discussed in Refs. [3] and [6]. Typical experimental result appear in Fig. 1.7.

The "crossover" peaks that occur when two of the levels are rigorously degenerate can be enormously useful. Such cases occur whenever one of the interacting levels has angular momentum greater that $\frac{1}{2}$. Those resonances are the origin of "polarization spectroscopy" [14–16].

Four clear examples of this type of spectroscopy are the transitions between $^1S_0$ and $^1P_1$ states diagrammed in Fig. 3.5. A right circularly polarized pump wave propagating along the positive z axis couples the states indicated by the double arrow in Figs. 3.5a and 3.5c. A pump linearly polarized along x couples the indicated levels in Figs. 3.5b and 3.5c. As before, the populations of the resonant velocity groups are perturbed not only in the coupled levels, but also in the others through collisional and decay processes. To treat the resulting effects properly, it is necessary to obtain the steady-state solutions of the appropriate 4 × 4 density matrix.

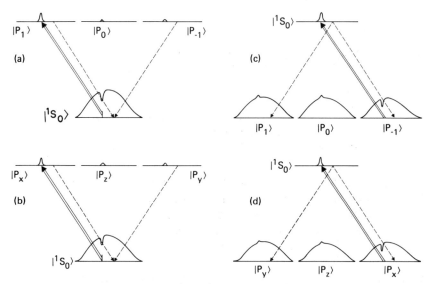

Fig. 3.5 Polarization spectroscopy in $^1S_0 \rightarrow {^1P_1}$ transitions. (a, b) The level systems are termed "V" systems and optical pumping plays little role. The pump wave is indicated by a double arrow while the probe wave polarization is such that it senses the difference in population on the transitions indicated by dashed lines. (a, c) The basis set appropriate for a circular polarized pump; (b, c) the basis best for linear polarization. In each case, the absorption on one of the coupled transitions is predominantly affected by the pump while the polarization spectroscopy signal depends upon the difference in population on the two transitions coupled by the probe beam. (c, d) Optical pumping in the "Λ" level system increases the polarization signal.

In polarization spectroscopy, the counterpropagating probe beam is linearly polarized. The pump can be either linearly or circularly polarized as above. If the pump is linearly polarized, it makes an angle of 45° with respect to the probe field $\mathbf{E}_-$. The nonlinear signal amplitude is polarized perpen-

3.2 Crossover Resonances and Polarization Spectroscopy

dicular to the initial probe field direction. The probe senses the population difference of two degenerate transitions as indicated by the dashed arrows. A polarizer properly oriented as in Fig. 3.6 can select this amplitude which results from the influence of the pump wave on the medium.

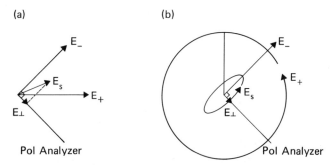

Fig. 3.6 The pump, probe, and signal polarizations in polarization spectroscopy. In each case, the probe wave (denoted E_-) is linearly polarized. (a) The pump wave ($\mathbf{E}_+$) is linearly polarized at 45° to the probe. The wave transmitted through the sample has the direction $\mathbf{E}_s$ as a result of the optically induced anisotropy of the medium which attenuates and phase shifts the component of $\mathbf{E}_-$ that is parallel to $\mathbf{E}_+$ by amounts different from the component perpendicular. The detected intensity is selected by the polarization analyzer and corresponds to $\mathbf{E}_\perp$, the component of $\mathbf{E}_s$ perpendicular to $\mathbf{E}_-$. Thus, the amplitude reaching the detector is $\mathbf{E}_\perp$. (b) The pump is circularly polarized and the induced anisotropies affect the circular components of $\mathbf{E}_-$ differently. The transmitted amplitude $\mathbf{E}_s$ is elliptically polarized on-resonance, and the polarization analyzer selects the linear component $\mathbf{E}_\perp$ for detection. In "biased" or optical-heterodyne-detected schemes, both the phase and amplitude of $\mathbf{E}_\perp$ are accessible.

The characteristics of this signal can be most easily understood in the case of Figs. 3.5b and 3.6a in the absence of collisions. The linearly polarized pump beam partly saturates the $|^1S_0\rangle - |P_x\rangle$ transition producing a difference in the velocity distributions between the upper and lower states as in Eq. (3.1.3). The probe amplitude can be resolved into components

$$\mathbf{E}_- = (E_-/\sqrt{2})(\hat{x} + \hat{y}) \tag{3.2.3}$$

and the complex amplitude of the x-polarized component is altered as in (3.1.6). The relevant nonlinear susceptibility tensor element is $\chi^{(3)}_{1111}$ in (3.1.11).

The difference in the velocity distributions for the $|^1S_0\rangle - |P_y\rangle$ level system is perturbed less by the pump field since no atoms are promoted into the $|P_y\rangle$ state,

$$g_{D0y}(v) = \{-1\}\frac{e^{-v^2/v_0^2}}{\sqrt{\pi}v_0}\left(1 - \frac{|\chi_+|^2 T_2 T_b/2}{(\Omega - \omega + kv)^2 + T_2^{-2} + |\chi_+|^2 T_b/T_2}\right). \tag{3.2.4}$$

The complex amplitude of the y-polarized probe component is thus altered less. The relevant nonlinear susceptibility tensor element is

$$\chi^{(3)}_{2112}(-\omega, \omega, -\omega, \omega) = \frac{n^2(\omega)}{48\hbar^3}|\mu_{ab}|^4 \frac{\mathcal{N}\{-1\}}{\sqrt{\pi\Omega_D}} \frac{T_b e^{-(\Omega-\omega)^2/\Omega_D^2}}{\Omega - \omega - iT_2^{-1}} \quad (3.2.5)$$

and the y-polarized part of the signal amplitude is

$$E_{sy} = \frac{24\pi i\,\omega l}{n(\omega)\,c} \chi^{(3)}_{2112}|E_+|^2 E_{-y} e^{-i(\omega t - kz)}$$

$$= \{-1\} \frac{-i\pi\omega\ln(\omega)\mathcal{N}}{2c\sqrt{\pi\Omega_D}}|\mu_{ab}|^4 \frac{T_b e^{-(\Omega-\omega)^2/\Omega_D^2}}{\Omega - \omega - iT_2^{-1}}|E_+|^2 E_{-y} e^{-i(\omega t - kz)}. \quad (3.2.6)$$

The anisotropy in the pumped sample thus alters the polarization state of the probe. On resonance, the transmitted probe is linearly polarized in the direction indicated by E_s in Fig. 3.6a.

The detected amplitude corresponds to the projection of E_s on the axis perpendicular to the initial probe polarization

$$\mathbf{E}_\perp = \left\{\frac{(\hat{x} - \hat{y})}{\sqrt{2}} \cdot \mathbf{E}_s\right\} \frac{\hat{x} - \hat{y}}{\sqrt{2}}$$

$$= -\frac{12\pi i\,\omega l}{n(\omega)\,c}\{\chi^{(3)}_{1111} - \chi^{(3)}_{2112}\}|E_+|^2|E_-|\frac{\hat{x} - \hat{y}}{\sqrt{2}}e^{-i(\omega t - kz)}$$

$$= \{-1\}\frac{-i\pi\omega n(\omega)l\mathcal{N}}{2c\sqrt{\pi\Omega_D}}|\mu_{ab}|^4 \frac{T_b e^{-(\Omega-\omega)^2/\Omega_D^2}}{\Omega - \omega - iT_2^{-1}}$$

$$\times |E_+|^2|E_-|\frac{\hat{x} - \hat{y}}{\sqrt{2}}e^{-i(\omega t - kz)}, \quad (3.2.7)$$

where the last two equations result from substituting (2.5.9), (2.9.7), (3.1.6), and (3.2.6). In the "background free" geometry, the intensity corresponding to the absolute square of $\mathbf{E}_\perp$ is detected. In other variations of the polarization spectroscopy, the electrical signal may correspond to the real or imaginary part of $\mathbf{E}_\perp$. The real part of $\mathbf{E}_\perp$ (or the imaginary part of $\chi^{(3)}_{1111} - \chi^{(3)}_{2112}$) corresponds to an intensity dependant dichroism, while the imaginary part of $E_\perp$ (real part of $\chi^{(3)}_{1111} - \chi_{2112}$) corresponds to birefringence. Clearly, a multilevel system is required for the existence of such anisotropies.

In general, the kind of anisotropy induced by a linearly polarized pump beam is termed "alignment" [16, 17]. Collisional effects can relax the alignment by equalizing the populations of states $|b\rangle$ and $|c\rangle$. The result is that

3.2 Crossover Resonances and Polarization Spectroscopy

the factor T_b in (3.2.7) must be replaced by $(T_b^{-1} + \tau_a^{-1})^{-1}$, where τ_a^{-1} is the collisional alignment relaxation rate.

The argument for the case shown in Fig. 3.5a is very similar except that the probe must be decomposed into circularly polarized components. The detected amplitude is

$$\mathbf{E}_\perp = \frac{12\sqrt{\pi}\,\omega l}{n(\omega)\,c} \{\chi^{(3)}_{2121}(-\omega,\omega,-\omega,\omega)$$

$$- \chi^{(3)}_{2211}(-\omega,\omega,-\omega,\omega)\}|E_+|^2|E_-|\frac{\hat{x}-\hat{y}}{\sqrt{2}} e^{-i(\omega t - kz)}$$

$$= \{-1\}\frac{\pi\omega n(\omega) l \mathcal{N}}{2c\sqrt{\pi}\Omega_D}|\mu_{ab}|^4 \frac{T_b e^{-(\Omega-\omega^2)/\Omega_0^2}}{\Omega - \omega - T_2^{-1}}|E_+|^2|E_-|\frac{\hat{x}-\hat{y}}{\sqrt{2}} e^{-i(\omega t - kz)} \quad (3.2.8)$$

which is 90° out of phase with that in (3.2.7). Collisions affect the angular momentum eigenstates differently from the x, y, and z orbitals. The kind of anisotropy induced by a circularly polarized pump beam is termed "orientation," and the orientation relaxation rate is slower in general than the alignment rate [16, 17]. When collisions are included, $(T_b^{-1} + \tau_0^{-1})^{-1}$ replaces T_b in (3.2.8).

The ratio of the signal amplitudes obtained with linear and circular pumps can indicate the angular momenta of the upper and lower levels. According to Saikan [18],

$$\frac{\chi_{1212} - \chi_{1122}}{\chi_{1111} - \chi_{1221}} = \begin{cases} \dfrac{5}{16J^2 + 16J - 7}, & \Delta J = 0, \\[6pt] \dfrac{10J^2 - 5rJ - 5}{18J^2 - 4rJ + 7}, & \Delta J = \pm 1, \end{cases} \quad (3.2.9)$$

where

$$r = \frac{\Delta J}{|\Delta J|}\frac{T_a^{-1} - T_b^{-1}}{T_a^{-1} + T_b^{-1}}$$

and $\chi_{1212} - \chi_{1122} = 0$ for $J = 0 \leftrightarrow J = 0$ transitions and $\chi_{1111} - \chi_{1221} = 0$ for $0 \leftrightarrow 0$ and $\frac{1}{2} \leftrightarrow \frac{1}{2}$ transitions.

Optical pumping must be included in the discussion of the cases in Figs. 3.5c and 3.5d. In these level systems, the pump creates an alignment or orientation of the ground state which may relax *only* as a result of collisions. The correct forms for the steady-state signal amplitudes then contain factors of τ_a and τ_0 in (3.2.7) and (3.2.8), respectively, instead of T_b. The polarization spectroscopy signals are much larger in this case. Generally, polarization spectroscopy is most useful when the lower level has a higher angular momentum and optical pumping of the ground state can enhance the detected anisotropy. Except for collisional effects, the linewidths extrapolated to zero intensity are the same as for saturated absorption. Complications also exist due to coherence and standing wave effects but need not be discussed here [16, 19].

3.3 COUPLED DOPPLER-BROADENED TRANSITIONS

Additional effects occur when the medium can support two transitions with different frequencies [3, 4, 6]. Three possible energy level systems of this sort appear in Fig. 3.7. The extra complexity results from the increased importance of "coherence effects" which must be considered even when there is no linear absorption. The coherent Raman effects and two-photon absorption resonances discussed in Chapters 4 and 5 result entirely from these effects. In the context of saturated absorption spectroscopy, these effects are important because they can interfere constructively and destructively with the resonances resulting from the population anomalies due to hole burning. All these effects correspond formally to the same order of nonlinearity in the cross-saturation case. The magnitude, frequencies, and line shapes of the resulting resonances can be altered in significant ways. In light of the excellent reviews of this topic, there is no need to do more here than to present the most salient features [4, 6, 8].

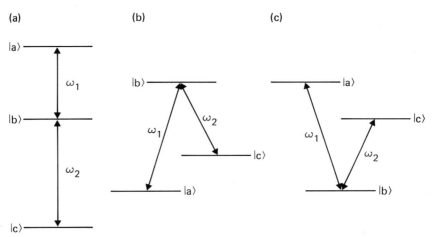

Fig. 3.7 Three-level systems with coupled transitions. The level splittings are assumed large enough that the Doppler profiles do not overlap. (a) The "stretched" level system and (b), the "Λ" and "V" folded systems, respectively. In certain circumstances, reduced or zero Doppler width resonances can be obtained in all of the indicated one- and two-photon transitions.

3.3 Coupled Doppler-Broadened Transitions

The natural context for a calculation including population and coherence effects is the 3×3 density matrix for the three-level system. At low intensities, the results for all three strongly Doppler-broadened cases can be summarized in terms of the third-order nonlinear susceptibility [6],

$$\chi^{(3)}_{ijkl}(-\omega_2, \omega_1, -\omega_1, \omega_2) = \frac{2\sqrt{\pi}\langle c|\mu_i|b\rangle\langle b|\mu_j|c\rangle\langle b|\mu_k|a\rangle\langle a|\mu_l|b\rangle}{\hbar^3 \Delta\Omega_{ab}}$$

$$\times \exp\left[-\left(\frac{\omega_1 - \Omega_{ab}}{\Delta\Omega_{ab}}\right)^2\right]\left\{\left(\frac{N_b}{g_b} - \frac{N_a}{g_a}\right)\Gamma_{bb}^{-1}\frac{\varepsilon'\Omega' + i\Gamma_B}{\Omega'^2 + \Gamma_B^2}\right.$$

$$+ \left(\frac{N_b}{g_b} - \frac{N_a}{g_a}\right)\frac{\varepsilon'\Omega'(\Gamma_B + \Gamma_N) + i(\Gamma_B\Gamma_N - \Omega'^2)}{(\Omega'^2 + \Gamma_B^2)(\Omega'^2 + \Gamma_N^2)}$$

$$\times U_0\left(-\frac{\varepsilon k_1}{\varepsilon' k_2}\right)\left[\frac{\omega_2}{\omega_1} - \frac{\omega_2 - \omega_1}{\omega_1}U_0(\omega_1 - \omega_2)\right]$$

$$+ \left(\frac{N_b}{g_b} - \frac{N_c}{g_c}\right)\left(\frac{\omega_1 - \omega_2}{\omega_1}\right)\frac{2\varepsilon'\Omega'\Gamma_N + i(\Gamma_N^2 - \Omega^2)}{(\Omega'^2 + \Gamma_N^2)^2}$$

$$\left. \times U_0\left(-\frac{\varepsilon' k_2 + \varepsilon k_1}{\varepsilon' k_2}\right)\right\}, \qquad (3.3.1)$$

where

$$\Omega' = \omega_2 - \Omega_{bc} - (k_2/k_1)(\omega_1 - \Omega_{ab}), \qquad \Gamma_B = \Gamma_{bc} + |k_2/k_1|\Gamma_{ab},$$

$$\Gamma_N = \Gamma_B - (\Gamma_{ab} + \Gamma_{bc} - \Gamma_{ac})\left[\frac{\omega_2}{\omega_1} - \frac{\omega_2 - \omega_1}{\omega_1}U_0(\omega_2 - \omega_1)\right],$$

and

$$U_0(x) = \begin{cases} 0, & x \leq 0, \\ 1, & x > 1, \end{cases} \qquad \varepsilon' = \frac{E_c - E_b}{|E_c - E_b|}, \qquad \varepsilon = \frac{E_b - E_a}{|E_b - E_a|}.$$

In deriving (3.3.1), all waves are assumed to propagate along the z axis. A positive value of k_1 or k_2 implies propagation in the positive direction while a negative value of k_1 or k_2 implies that the wave propagation in the negative direction. The magnitudes of the wave vectors are related to the frequency by the index of refraction of the gaseous medium

$$|k_1| = n(\omega_1)\omega_1/c, \qquad |k_2| = n(\omega_2)\omega_2/c.$$

For weak enough transitions, the indices of refraction can be considered equal to unity, and some simplifications in (3.3.1) result. Also, N_α/g_α is the ratio of the population density of level α to its degeneracy. The Doppler width for the Ω_{ab} transition is

$$\Delta\Omega_{ab} = (v_0/c)\Omega_{ab},$$

and the population decay rate of the intermediate state $|b\rangle$ is Γ_{bb}. All frequencies other than Ω' are assumed positive.

Equation (3.3.1) has a number of unusual implications which can be seen most clearly for the case when $n(\omega) = 1$ and where pure dephasing is absent and thus $\Gamma_{ab} = (\Gamma_{aa} + \Gamma_{bb})/2$, etc. First, resonances do not appear at the actual transition frequencies but rather when the laser frequencies obey

$$\omega_2 = \Omega_{bc} \pm (\Omega_{bc}/\Omega_{ab})(\omega_1 - \Omega_{ab}), \qquad (3.3.2)$$

where the plus sign is taken for copropagating beams and the minus sign for counterpropagating waves. For a given value of ω_1, the values of ω_2 at which resonance occurs are displaced equally from Ω_{bc} for the two propagation directions. In deriving (3.3.2), the detunings from resonance were assumed much smaller than the level separations.

A second unusual feature is that the heights and widths of the resonances depend upon the propagation directions [4]. For the folded transition schemes of Figs. 3.7b and 3.7c with $\omega_2/\omega_1 \approx 1$ and co-propagating beams, the linewidths of the resonances are insensitive to the linewidth (i.e., the decay rate Γ_{bb}) of the common level $|b\rangle$,

$$\Gamma_{\rightrightarrows} = \begin{cases} \dfrac{\omega_2}{2\omega_1}(\Gamma_{aa} + \Gamma_{cc}) + \dfrac{1}{2}\left(1 - \dfrac{\omega_2}{\omega_1}\right)(\Gamma_{bb} + \Gamma_{cc}), & \omega_2 < \omega_1, \\ \dfrac{1}{2}(\Gamma_{aa} + \Gamma_{cc}) + \dfrac{1}{2}\left(\dfrac{\omega_2}{\omega_1} - 1\right)(\Gamma_{aa} + \Gamma_{bb}), & \omega_2 > \omega_1, \end{cases} \qquad (3.3.3)$$

while for counterpropagating beams, that rate can be the main contribution,

$$\Gamma_{\rightleftarrows} = \tfrac{1}{2}(\Gamma_{cc} + \Gamma_{bb}) + (\omega_2/2\omega_1)(\Gamma_{aa} + \Gamma_{bb}). \qquad (3.3.4)$$

For the copropagating beams, the contributions of Raman and stepwise transition processes add coherently to the radiated amplitude, while only the stepwise process contributes to the counterpropagating wave. Thus, the resonance obtained for copropagating beams is narrower, larger in area, and with a much higher maximum. Very similar considerations apply for the stretched-level diagram in Fig. 3.6a, but in this case, the roles of copropagating and counterpropagating beams are reversed, with narrower and stronger resonance occurring for oppositely directed beams. These asymmetrical effects reflect the importance of correctly averaging over the random distribution of thermal velocities in a gas. For some configurations, molecules with a wide range of velocities can all contribute to the resonance, but for others only a narrow range is effective. Experimental results verifying these asymmetries appear in Fig. 3.8.

3.3 Coupled Doppler-Broadened Transitions

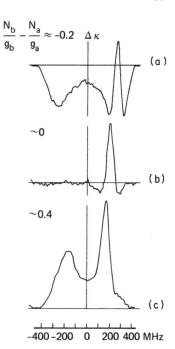

Fig. 3.8 Saturation spectra in the "V"-type folded three-level system Ne:$3s_2-2p_4-2s_2$. The pump waves at 0.6328 nm propagated in opposite directions through the sample which was probed at 1.15 μm. Thus, the spectra for copropagating and counterpropagating waves are superimposed. (b) Conditions in the discharge have been adjusted to make the sample transparent at the pump frequency. The remaining peak is due to Raman- type transitions and shows the narrowing expected for copropagating pump and probe. (a, c) The wider resonances are due to "stepwise" transitions with counterpropagating beams. The "stepwise" peak is positive or negative depending upon the sign of the effective population differences. These line shapes are substantially as predicted by Eq. (3.3.1) (from Ref. [1]).

Another aspect of (3.3.1) is the fact that the nonlinear resonances exist even when the populations are equalized and linear absorption vanishes. In that case, the resonant Raman and two-photon effects are observed uncomplicated by stepwise transition. These effects are discussed more fully in Sections 3.4 and 3.5.

Perhaps the most important application of the narrow resonances occurring for the folded-level diagram of Fig. 3.6b is the optically pumped three-level laser such as the I_2, Na_2, and S_2 lasers [20]. If neither the pumping beam nor the internal laser field itself is too strong, the gain and loss coefficients of these media are correctly described by (3.3.1). In a ring resonator, unidirectional laser action can be obtained as a result of the gain asymmetry. Laser action can take place only over a narrow range of frequencies closely correlated with the frequency of the pump, thus facilitating the transfer of frequency stability from one spectral region to another. At higher intensities, the gain bands split and broaden as the result of saturation and the Autler–Townes effect. Such phenomena cannot be correctly described in terms of a third-order nonlinear susceptibility. Various authors have invoked higher-order approximations and even obtained analytical formulas for the line-shape functions [4, 8].

3.4 EXPERIMENTAL METHODS OF SATURATION SPECTROSCOPY IN GASES

The earliest saturation spectroscopy experiments were performed upon the gain media of cw gas lasers. The output power of a single-mode gas laser can show a decided dip—the famous Lamb dip—when the output frequency corresponds to the transition frequency of atoms at rest [21]. The output frequency is typically tuned by mounting one of the laser mirrors on a piezoelectric translator and scanning the cavity length over the gain profile. The frequency of the laser can be locked electronically to the center of the dip, and this technique continues to be used in some frequency stabilized lasers intended for metrology.

The next significant innovation was the incorporation of a sample cell inside the laser cavity. Since oscillation of the laser was sustained by the gain of the laser medium, the sample cell could be operated at lower pressure or lower discharge current. The Lamb dips that appeared in the laser output were narrower and were displaced from those of the gain medium as the result of the pressure shift. Javan et al. devised a means of detecting the nonlinear resonances by monitoring the side fluorescence of the sample cell at a wavelength produced by a transition coupled to that being probed at the laser frequency. This "saturated fluorescence technique" monitors the population of the upper level and avoids some of the noise superimposed on the laser output [22].

When the sample cell is filled with an absorbing medium, the "Lamb dips" in the laser output are "inverted" as a result of the reduction of the intracavity loss that results from the saturation of the absorption [23]. This inverted Lamb dip technique allowed a wider variety of molecular species to be studied even with fixed frequency lasers. The spectroscopic resolution increased to previously unimaginable levels. A narrow resonance in CH_4 observed by Barger and Hall using this technique [2] is shown in Fig. 3.9. Dips in fluorescence can also be observed in absorbing samples by the Javan technique.

The inverted Lamb dip method retains the important advantages of simplicity and reproducibility. It continues to be employed in frequency stabilized lasers for precision traveling Michelson wavelength meters and in reference oscillators for long-term stabilization of tunable sources. The most common absorbing species is perhaps $^{129}I_2$ which has a relatively strong line near the 6.33 nm ^{3}He–^{20}Ne laser line. Natural iodine can also be used, but the resonances are weaker. The best stabilization has been achieved with methane and the HeNe 3.39-μm line. Two lasers of the same design with the same sample pressure and intracavity power level produce stabilized frequencies within a few kilohertz of one another using the inverted Lamb dip phenomenon.

3.4 Experimental Methods of Saturation Spectroscopy in Gases

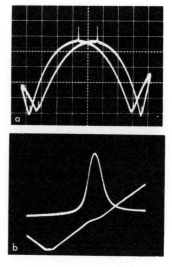

Fig. 3.9 Inverted Lamb dip due to the $F_1^{(2)}$ component of the P(7) line of the v_3 band of CH_4 at 3.39 μm. The methane cell was inside the cavity of a HeNe laser, and when the cavity was tuned to the center of the absorption line, the reduction of intracavity loss due to the Lamb dip effect caused an increase in laser power. (a) The 400-kHz-wide inverted dip superimposed on the Doppler-broadened gain band of the laser. (b) The resonance in an expanded scale along with a plot of laser output frequency as a function of mirror displacement. The narrow saturated dispersion resonance of the intracavity methane cell causes the tuning curve to flatten in the region of the inverted Lamb dip (from Ref. [2]).

One disadvantage of the inverted Lamb dip technique is that the sample absorption must be low enough that laser action takes place. Another is the difficulty of obtaining quantitative measurements. Near threshold, small changes in intracavity gain and loss produce large changes in the laser output power. Thus, it is difficult to measure the level of saturation. The dispersion associated with the decreased absorption alters the effective length of the resonator, pulling the laser frequency toward the center of the inverted Lamb dip. This saturated dispersion effect tends to increase the frequency stability of the laser but makes measurements of the line width of the nonlinear resonance almost impossible. A wise experimentalist perturbs the design of an operable laser as little as possible and does as much physics as possible outside the cavity.

The extracavity techniques of saturation spectroscopy are complicated by the fact that an output beam reflected back into the laser cavity tends to destabilize the oscillation frequency. Some sort of isolation is necessary between the laser and experiment. In the "old-fashioned" Hänsch–Borde arrangement diagrammed in Fig. 3.10a, the necessary isolation is arranged by allowing the pump and probe beams to cross at a small angle in the sample. Rudimentary spatial filtering to reject the returning probe beam can be achieved by properly positioning the mirror M2. The finite angle θ between the interacting beams introduces a residual linear Doppler width of roughly $\theta \Omega_D$ in the resonance line width. The interaction length of the beams is also limited. The nonlinear resonances are conveniently detected by chopping the pump beam and detecting the resulting modulation of the transmitted probe intensity [24].

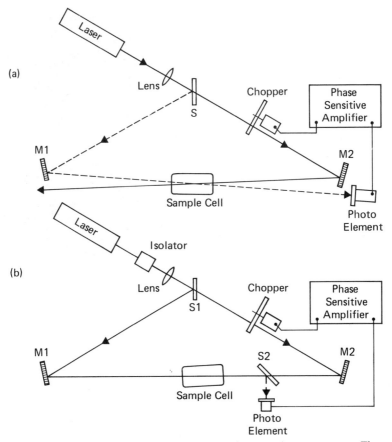

Fig. 3.10 Two versions of the Hänsch–Borde extracavity saturation spectrometer. The probe beam is shown as a dashed line while the pump beam is shown as a solid line. The change in probe transmission due to the pump is detected by chopping the pump wave and employing phase sensitive detection. (a) In the original configuration, a small angle between the beams prevents feedback into the laser cavity with the consequent disruption of oscillation. (b) The preferred geometry employs an optical isolator such as a Faraday rotator to prevent feedback and has the two beams exactly collinear. The signal is then extracted by a second beam splitter at S2 which directs part of the probe beam into a photodetector. The lens shown in each apparatus ensures that the beams are collimated in the interaction region.

The Hänsch–Borde configuration can be employed with truly counterpropagating beams as in Fig. 3.10b, if some means of isolation can be devised. For pulsed lasers, isolation can be simply a matter of positioning the experiment so far from the laser that reflections do not have time to return before oscillation terminates. Such pulsed lasers show relatively poor pulse-to-

3.4 Experimental Methods of Saturation Spectroscopy in Gases

pulse intensity and frequency stability, necessitating sophisticated signal processing. Another simple isolator consists of an absorbing filter that attenuates the transmitted laser beam by a factor of T. The returning beam is consequently a factor of T^2 weaker than it would be in the absence of the filter, and it is often possible to retain sufficient intensity for an experiment while reducing feedback into the laser cavity to an acceptable level.

A more sophisticated isolator employs the Faraday nonreciprocal polarization rotation of light transmitted through a material in a longitudinal magnetic field. The isolator consists of two polarizers, a polarization rotating medium, and a magnet. The polarizers are oriented to minimize the transmission of light reflected back into the laser cavity, and the magnetic field is adjusted to maximize transmission in the forward direction. While isolation can be achieved with rather small rotation angles, optimum power transmission requires a rotation of 45°. Such rotations can be obtained for visible wavelengths with a terbium gallium garnet rotator medium and 4-kG permanent magnets or with FR5 rotator glass and a solenoid electromagnet.

An even more sophisticated isolator employs an accoustooptic frequency shifter/deflector to direct the laser beams into the experimental region. Return beams are diffracted away from the laser axis and frequency shifted further from the laser frequency by the traveling sound wave in the medium. Such isolators can be engineered to counteract intensity and frequency instabilities in the laser as well as to provide isolation.

If the experiment incorporates no optical elements which modify the polarizations of the interacting beams, adequate isolation can sometimes be obtained with a polarizer and a quarter-wave plate. Since helicity is preserved upon reflection, the circularly polarized returning wave has the opposite polarization to the input wave and is rejected by the polarizer.

A highly evolved Hänsch-type saturation spectrometer appears in Fig. 3.11. The laser output is frequency shifted by an acoustooptic isolator and monitored with detector D1. The power in the sample can be adjusted without altering the laser parameters by varying the acoustic wave intensity. The beam is spatially filtered and expanded by the telescope T and then split into pump and probe beams by beam splitter B2. The pump beam is conventionally made stronger then the probe. An additional beam splitter B3 reflects a reference beam which monitors the linear absorption of the sample and is detected at D2. The pump and probe beams are directed into the sample by mirrors M2 and M3. Great care is necessary to ensure that they are indeed exactly counterpropagating, but the reflected fringe patterns from B1 and B2 can be employed diagnostically. The pump beam is chopped as shown, and the transmitted probe beam detected at D3. The signals from the three optical detectors, the chopper position, acoustic wave intensity, laser frequency, etc., are logged simultaneously by minicomputer for later analysis.

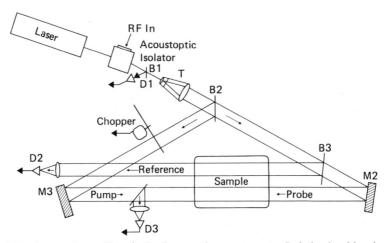

Fig. 3.11 A more elegant Hänsch–Borde saturation spectrometer. Isolation is achieved using an accoustoptic modulator. Detector D1 monitors the laser power; the telescope T expands and collimates the beams to reduce transit time broadening. Beam splitter B2 separates the pump from the probe beams, while B3 splits off a reference beam which measures the unsaturated transmission of the sample. The pump and probe beams are superimposed with mirrors M2 and M3, with the saturation spectrum appearing as a modulation at detector D3.

Analog electronics is also employed to provide real-time diagnostics as to the operation of the apparatus.

A somewhat simpler means of achieving the same sort of precision is indicated in Fig. 3.12 [52]. Here the standing wave field necessary for saturation spectroscopy is achieved by means of cat's eye expanded field retroreflectors. Laser power is monitored by detector D1, and detector D2 measures both the nonlinear signal and sample absorption. The sensitivity is enhanced by modulating the frequency of the laser electrooptically. The

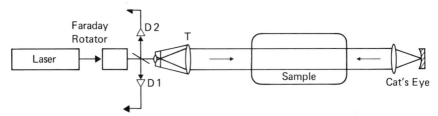

Fig. 3.12 A simple standing-wave extracavity saturation spectrometer incorporating a beam-expanding telescope and an expanded-field cat's eye retroflector to ensure that the pump and probe propagate in exactly opposite directions. Detector D1 monitors the laser power while D2 detects the saturated absorption signal. Slight frequency modulation of the laser facilitates signal detection, but produces a derivative line shape.

3.4 Experimental Methods of Saturation Spectroscopy in Gases

modulation of the intensity at D2 reflects the derivative of the saturated absorption line shape. Again data logging is by computer. The high quality of the data obtained from such a system when care is taken to eliminate every complicating spurious phenomenon is illustrated in Fig. 3.19.

The nonlinear resonances can also be detected by saturated fluorescence in extracavity configurations. A system in which the pump and probe beams are chopped by a single chopper wheel with three sets of openings is illustrated in Fig. 3.13. The pump is chopped at rate F_1, the probe at F_2, and the modulation of the fluorescence detected at frequency $F_1 + F_2$ or $F_1 - F_2$ which is produced at the reference detector by the third set of openings [1–3]. If the sample is excited by a stable discharge, the same modulation frequency appears on the discharge current as a result of the optogalvanic effect.

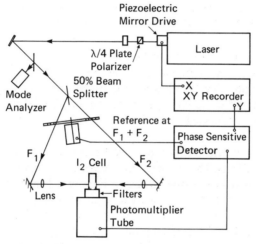

Fig. 3.13 Detection of saturation spectroscopy resonances by intermodulated fluorescence. In this and other saturated fluorescence techniques, the output signal is proportional to the population of the excited state which is less when the counterpropagating beams are interacting with the same velocity group than when they interact with separate velocity groups. In the intermodulated fluorescence version, the two beams are chopped at different frequencies and the component of the fluorescence oscillating at the sum frequency is identified electronically (from Ref. [2]).

Some of the advantages of the intracavity techniques for studying weak transitions with low power lasers can be retained by enclosing the sample cell in a second Fabry–Perot resonator. An experiment of this type in which the build-up cavity is engineered to produce a large-diameter region with nearly plane wavefronts in the region of the sample is diagrammed in Fig. 3.14. The length of the build-up cavity must be servoed to the laser

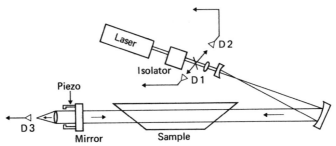

Fig. 3.14 A saturation spectrometer in which the amplitude of the interacting beams is enhanced in an optical resonator. The length of the cavity containing the sample must be served piezoelectrically to the output wavelength of the laser. On resonance, the intensity of the beam in the sample is enhanced by a factor equal to the cavity finesse which can be as large as 1000. The saturation resonances perturb the cavity producing changes in the reflected and transmitted intensities at detectors D2 and D3. The cavity shown employs three mirrors to maximize the beam diameter and minimize the phase front curvature in the sample region. Carefully selected lenses are necessary to match the transverse modes of the laser and cavity.

frequency with electronics that can follow any necessary modulations. Inside the build-up cavity, the light intensity can be several hundred times larger than in a corresponding two-beam experiment. The mode quality is also generally better, and small changes in the loss or phase shift produce larger variations in the transmitted and reflected intensities monitored by detectors D2 and D3.

Other interferometric techniques have been employed to measure saturated dispersion effects. The simplest such technique uses a Jamin interferometer with the sample cell in one arm as in Fig. 3.15. The intensity and phase of the wave in the other arm is adjusted to give a local oscillator amplitude E_{LO} at the detector that adds with the nonlinearly radiated amplitude E_s in quadrature [see Eq. (2.9.13)]. In practice, achieving this condition requires servo electronics. If the phase and amplitude of the local oscillator are incorrect, distorted lineshapes appear [25].

Polarization spectroscopy formally resembles the nonlinear interferometry techniques, but is simpler in practice. It achieves its increased sensitivity partly by rejecting most of the laser noise that reaches the detector in other saturation techniques and partly by increasing the signal amplitude by means of optical pumping. The polarization spectroscopy apparatus diagrammed in Fig. 3.16 superficially resembles the Hänsch–Borde saturation spectrometer in Fig. 3.10b. The linear polarizer at P ensures that the probe beam is highly linearly polarized in a direction likely to remain unmodified by the sample windows and beam splitter B2. The portion of the transmitted probe reflected by this beam splitter passes through a variable wave plate (Babinet–Soleil compensator) at VW which, in the simplest schemes, cancels

3.4 Experimental Methods of Saturation Spectroscopy in Gases

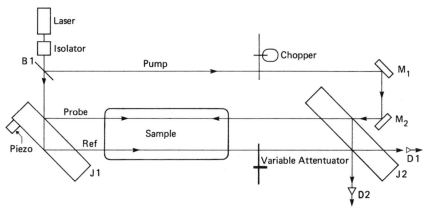

Fig. 3.15 A saturation spectrometer employing a Jamin interferometer. The pump and probe beams interact with the sample along one beam line. A reference beam is also transmitted through the sample, interfering with the transmitted probe at interferometer plate J2. The relative intensities of the two waves can be adjusted with the variable attenuator, while their phase difference can be altered by tipping plate J1 with the indicated piezoelectric translator. Away from a saturation resonance, the sum of the transmitted probe and reference amplitude corresponds to the local oscillator amplitude E_{LO}. The relative intensity and phase of the transmitted probe and reference waves are adjusted to give a local oscillator phase which produces a modulation of the intensity at the detectors D1 and D2 that reflect the saturated absorption, saturated dispersion or some combination of the two (from Ref. [25]).

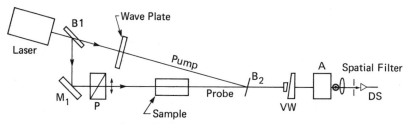

Fig. 3.16 An elementary polarization spectrometer. The probe beam is linearly polarized by the polarizer at P, while the pump can be either linearly or circularly polarized depending upon the wave plate setting. The two beams interact as they counterpropagate through the sample. The variable wave plate (Babinet Soleil compensator) at VW and the polarization analyzer at A can be adjusted to block the probe for off-resonant frequencies or to allow a local oscillator amplitude to reach the detector DS.

out the effects of birefringence in the optical system. A polarization analyzer A is adjusted to reject the transmitted probe when the pump beam is absent. Early workers emphasized the importance of achieving the maximum rejection; with care only 1 part in 10^8 of the probe is incident on detector DS [14].

The pump beam is either circularly polarized or linearly polarized at an angle of 45° with respect to the probe. The interactions of the sample with the light leads to a (linear or circular) birefringence and dichroism which alters the state of polarization of the probe. In the simplest polarization spectroscopy scheme, the intensity proportional to the component of the probe perpendicular to the initial probe polarization is transmitted through the analyzer at A and detected by DS. The signal produced is proportional to $E_\perp^2$ and Lorentzian in line shape, generally much weaker than the saturation spectroscopy signal (see Eqs. 3.2.7 and 3.2.8).

If the polarization analyzer is rotated a small amount, an additional amplitude E_{LO} proportional to the angle of rotation and either in phase or 180° out of phase with the probe amplitude is transmitted into the detector. The nonlinear signal intensity is proportional to $(2\,\mathrm{Re}\,E_{LO}^* \cdot E_\perp + E_\perp^2)$ where the first term can easily be made to dominate by setting $E_{LO} \gg E_\perp$. With a circularly polarized pump, the line shape of the first term is dispersionlike, while that of the second remains Lorentzian. If, however, the phase retardation of the variable wave plate VW is altered, the local oscillator amplitude transmitted through the analyzer A will be in quadrature with the probe amplitude. Positive and negative Lorentzian resonances result from the interference of $E_\perp$ and E_{LO}. These resonances are much larger than those obtained when the transmitted probe intensity is minimized. The exact local oscillator level producing the optimum sensitivity depends upon the characteristic of the laser and detection system, and will be discussed in Section 4.10.

Doppler-free laser induced dichroism and birefringence can also be detected electronically by employing a probe beam of mixed polarization and an analyzer capable of directing the separated components into two detectors. With a linearly polarized pump, the probe should be circularly polarized and vice versa. With a circularly polarized probe and a rotating analyzer, Delsart and Keller simultaneously recorded both phases and showed that the background pedestal due to atoms with collisionally altered velocities was absent [26]. The development of variations of these techniques in which the pump, probe, and analyzer polarizations are modulated seem to be limited only by imagination.

3.5 RAMSEY FRINGES IN SATURATION SPECTROSCOPY

The uncertainty principle limits the resolution of any spectroscopy experiment to the inverse of the time during which an atom interacts with the light. Consider the saturated fluorescence experiment illustrated in Fig. 3.17a. The atoms from an atomic beam cross a standing wave field with Gaussian

3.5 Ramsey Fringes in Saturation Spectroscopy

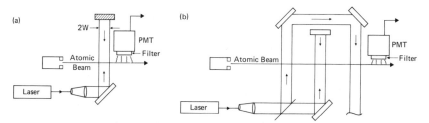

Fig. 3.17 Eliminating transit time broadening using Ramsey fringes in saturated fluorescence. (a) The minimum linewidth obtainable is inversely proportional to the time an atom requires to transit the beam in the single interaction region. (b) Interaction amplitudes for the three interaction regions interfere to produce "fringes" on the saturated fluorescence line shape with a width inversely proportional to the time the atoms require to move between beams.

profile of width $2w$, and the fluorescence that results is detected on the other side of the beam. In the rest frame of an atom, it experiences a pulse of radiation

$$E(t) \approx E_0 e^{-(vt/w)^2}$$

with Fourier components $E(\omega)$,

$$E(\omega) \approx E_0 e^{-(\omega w/2v)^2}.$$

When all other effects are absent, the linewidth of the nonlinear fluorescence resonance necessarily reflects the frequency width of the exciting pulse $\Delta\omega = 2v/w$. That width can be reduced by enlarging the beam diameter, but such a strategy is not an efficient allocation of limited laser power. Even if anamorphic optics have been employed to produce a highly elliptical laser spot, most of the radiation would have been wasted someplace where the atoms were not present. Moreover, the optical Stark effect would act to shift the energy levels of the atom during its entire passage through the beam.

The resolution of such an experiment is analogous to that of a telescope. In each case, the narrowest feature resolvable is related to the Fourier transform of the aperture function. In a telescope, the relevant aperture is that of the objective; in laser spectroscopy, it is that of the interaction region. Michelson showed that an interferometric technique can be used to enhance the resolution of a telescope without increasing the size of the objective. He mounted relatively small mirrors far apart and forced the waves reflected into the prime focus to interfere coherently. The result was a fringe pattern that gave the telescope a resolution equivalent to one with an aperture as large as the spacing between the small mirrors. Today, radio astronomers use this synthetic aperture technique to achieve a resolution equivalent to that of a telescope as large as the earth.

The Ramsey fringe technique improves spectroscopic resolution by forcing the interactions of an atom with separated beams to interfere coherently in the quantum mechanical sense [27]. The equivalent interaction region is as large as the space between the beams. Again, a fringe pattern results; when the center is identifiable, the resolution and precision of the experiment are even more greatly enhanced than the resolution of Michelson's telescope.

The applications of Ramsey's methods to Doppler-free multiphoton spectroscopy in Section 5.2 are easier to understand than the applications to saturated absorption, and the entire subject will be treated more fully in Section 6.5. Here we discuss the experimental methods necessary to observe Ramsey fringes in saturated absorption. Chebotayev was the first to point out that the random trajectories of individual atoms would tend to wash out the phase information necessary to produce observable fringes within the saturation resonances of an inhomogeneously broadened transition [13]. The fringes, however, can be made to appear when the beams are arranged to reverse the dephasing that results from the randomness of the trajectories. Thus, the Chebotayev–Ramsey fringe technique is analogous to the stimulated echo discussed in Section 6.4.

The necessary innovation is diagrammed in Fig. 3.17b. At least three interaction regions are required, instead of the two implied by the analogy with Michelson's stellar interferometer. In the first interaction region, atoms in the ground state are excited by a traveling light wave which results in an off-diagonal density matrix element of the form

$$\rho_{ab} \propto e^{ikz},$$

where z is the direction along the propagation direction of the light. As the atoms continue to precess at their own resonant frequencies as in Eq. (2.4.20), however, they arrive the second interaction region with essentially random axial positions z. The spatial component of the initial coherent ensemble has dephased even though the atoms retain information as to their individual phases. In terms of the vector model, the state of the ensemble has evolved from one with the **R** vectors for the individual atoms oriented along the 2 axis, as in Fig. 2.5 of one in which the vectors are uniformly distributed around the 1-2 plane.

In the second interaction region, the atoms encounter a standing wave. The atomic **R** vectors that entered the second interaction region at an angle of ϕ to the 2 axis, leave that region with an angle of $-\phi$. Since no momentum has been transferred from the standing wave field to the ensemble, the spatial Fourier components are unaltered.

As the atoms travel from the second interaction region to the third, the free precision *undoes* the phase shifts induced between the first and second regions so that at the third there is a component of the density matrix of the form $\rho_{ab} \propto e^{-ikz}$. The interaction with the beam propagating in the third inter-

3.5 Ramsey Fringes in Saturation Spectroscopy

action region in the $-z$ direction puts more atoms in the excited state $|b\rangle$ when the polarization interferes destructively with the optical field, and drains atoms from the excited state when the interference is constructive.

Performing the averages over position and axial velocity in the limit of large Doppler width yields a formula for the population difference which is monitored by the fluorescence,

$$\Delta w \propto \frac{\cos(2(\omega - \Omega)t_t + \Phi)}{(\omega - \Omega)^2 + \Gamma_{\text{eff}}^2}.$$

The time t_t appearing in the numerator is the time it takes an atom to propagate between interaction regions. In a real atomic beam, there is a distribution of such times. Averaging over these times washes out all but the central fringe. The phase factor in the argument of the cosine results from phase shifts between the beams in the different interaction regions at the plane $z = 0$, $(\omega - \Omega)$ is the detuning from resonance and Γ_{eff} is the "normal" width of the Lamb dip [see Eq. (6.5.8)]. To realize the full precision of the Ramsay fringe technique, it is necessary to set the phase Φ to zero or π.

In Berquist et al.'s implementation of this scheme shown in Fig. 3.18, standing waves exist in each interaction zone, but there are no additional complications [28]. The three beams are aligned parallel by cat's eye retro-

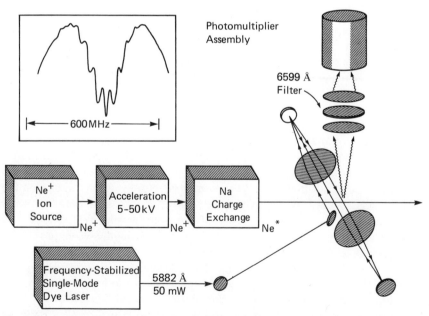

Fig. 3.18 An actual Ramsey fringe spectrometer. The large ellipses represent lenses used in the cat's-eye retroreflector optics and to collect fluorescence. The inset shows a saturated fluorescence resonance with the expected fringe structure (from Ref. [28]).

reflectors and the atomic beam is monoenergetic. The experimental traces in the inset clearly show the expected fringe structure within the Lamb dip in the detected fluorescence. An improved experiment of this type produced the narrowest observed optical resonance—1.7 kHz in atomic calcium [29].

3.6 THE LINE-SHAPE PROBLEM IN SATURATION SPECTROSCOPY

In spite of many years of effort, the present understanding of the line shapes observed in saturation spectroscopy is not complete [30]. Even when collisions and power broadening are absent and long-lived states interact with a perfectly uniform beam with plane wave fronts, the line shapes remain more complex than expected. Under these perfect conditions, the complexities result from relativity and quantum mechanics. After all, the saturation resonances result from atoms emitting a photon in one direction and absorbing one propagating in the opposite direction. Either event can occur before the other, but the atom recoils as the result of momentum transfer from the photon field. The result is a recoil splitting of the saturation resonance, the splitting having a magnitude of

$$\Delta\omega = \hbar\omega^2/Mc^2.$$

Slow-moving atoms remain in the interaction regions longer than faster atoms and hence give a larger contribution to the nonlinear signal. The quadratic or transverse Doppler effect predicted by special relativity implies that the more numerous fast moving atoms will exhibit different resonant frequencies even if their velocities are exactly parallel to the wave fronts. Such being the case, where is the center of the recoil split doublet with respect to the transition frequency of an atom truly at rest? [31].

When the intensity of the interacting beams is not uniform or the wave fronts are not plane, transit time effects alter the line shape. Time-dependant and time-independant optical Stark shifts lead to asymmetries and reduce the recoil doublet splitting. Collisions cause pure dephasing and thus broaden the resonances, but they also alter atomic trajectories without damping the oscillations. Weak velocity changing collisions broaden the saturation resonances by increasing the width of the hole in velocity space. Strong velocity changing collisions remove atoms entirely from the resonant velocity groups thus adding a "lifetime" broadening.

When optical pumping is significant, as in polarization spectroscopy, the lifetime of the holes burned in velocity space can be so long that collisions should broaden them to essentially the full Doppler width before decay. Narrow resonances, however, remain [53]. Finally, no laser is perfectly monochromatic. What effect do the inevitable phase and amplitude glitches have on the ensemble? Can the limit of zero power broadening be approached?

3.7 EXPERIMENTAL RESULTS IN SATURATION SPECTROSCOPY OF GASES

How much of its history does an ensemble remember? Could there perhaps be a better means of determining the center frequency of a Doppler-broadened transition than saturation spectroscopy?

3.7 EXPERIMENTAL RESULTS IN SATURATION SPECTROSCOPY OF GASES

It is not possible to adequately survey the results of a full decade of saturation spectroscopy by the world laser community in a few pages of text. Extensive but incomplete surveys are presented in one book and a number of excellent review articles [1–5]. There have been a great many papers describing aspects of saturation spectroscopy and verifying various aspects of the theory of these effects. We shall instead sample saturation measurements of atomic and molecular parameters. Among the parameters measured have been term values, rotational and vibrational constants, hyperfine constants, Zeeman and Stark coefficients, as well as collisional shift and broadening cross sections.

The progress in applying saturation techniques to probe fine details of molecular spectra is well illustrated by a continuing series of experiments on the components of the P(7) line of the v_3 band of CH_4 coincident with the 3.39 μm line of the helium–neon laser. An early experimental trace appears in Fig. 3.9. The narrow spike is an inverted Lamb dip. When a magnetic field was applied, the spike due to the F(2) line split into components corresponding to the Zeeman selection rules $M = \pm 1, 0$. The magnetic moment results from molecular rotation, with $g_J = +0.311 \pm 0.006$. At still higher resolution, the magnetic hyperfine structure in Fig. 3.19a appears. The center frequency

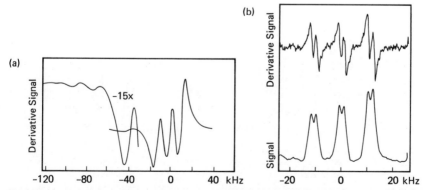

Fig. 3.19 Improving resolution in the saturated absorption spectra of CH_4 at 3.39 μm. Both traces correspond to the narrow spike in Fig. 3.9. (a) Three major magnetic hyperfine components along with some subsidiary structure. (b) At the higher resolution, those components are split by the recoil effect (from Ref. [7]).

for the transition was reported as 88,376.181627 (50) GHz. At the highest resolution, the hyperfine components are split by the recoil effect as in Fig. 3.19b. The hyperfine constants have been measured to 50 Hz in the ground state and 500 Hz in the excited state and are within four standard deviations of theoretical expectations.

Similar, though less extensive, programs have focused on a number of other molecules including NH_2D, CH_3F, SF_6, OsO_4, and I_2 [1]. Iodine is perhaps the most interesting as its rovibronic absorption spectrum spans the visible with millions of lines each with a distinct hyperfine structure. Two typical lines appear as resolved in saturated absorption in Fig. 3.20. Lines with odd values of J split into 21 components in seven groups, while those with even values of J split into 15 components [32]. A pair of nearly degenerate rovibronic lines lies within the gain band of the Ar^+ laser at 5145 Å. A recent derivative spectrum of the hyperfine structure of these lines, conventionally assigned as the P(13)R(15)43-0 of the X–B electronic band, is shown in Fig. 3.21. The crossover peaks are clearly visible and allow the separate determination of the hyperfine constants in the ground and excited states [33]. As many as six coupling constants are required to describe the splitting of each level. These constants vary with the rotational and vibrational quantum number with the magnetic couplings diverging at the dissociation limit.

In atomic spectroscopy, the most significant experiments have focused on hydrogen. Resolving the puzzles in the spectrum of this atom led to many advances in understanding. At each stage, improved resolution and improved precision challenged theoretical predictions. Hydrogen, however, is the lightest atom, with the largest fractional Doppler width. By removing the frequency uncertainty due to the Doppler width, saturation spectroscopy improved the precision of fundamental physical constants.

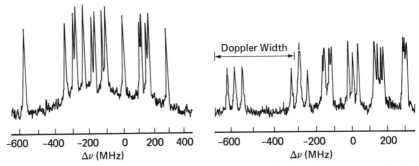

Fig. 3.20 Typical saturated absorption spectra of I_2 circa 1971. The splittings among the 15 or 21 hyperfine components reflect the electric quadrupole and spin-rotation interactions for the two spin-5/2 ^{127}I nuclei (from Ref. [32]).

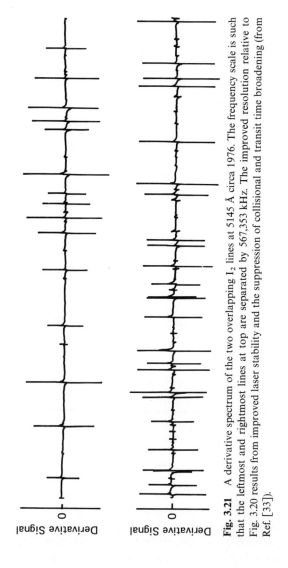

Fig. 3.21 A derivative spectrum of the two overlapping I$_2$ lines at 5145 Å circa 1976. The frequency scale is such that the leftmost and rightmost lines at top are separated by 567,353 kHz. The improved resolution relative to Fig. 3.20 results from improved laser stability and the suppression of collisional and transit time broadening (from Ref. [33]).

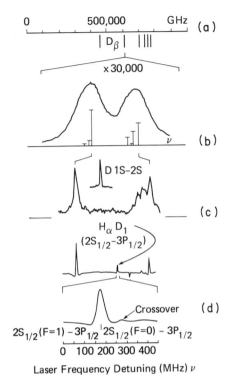

Fig. 3.22 The improving resolution of the spectroscopy of hydrogen. (a) The Balmer series as revealed by a simple spectrometer. (b) At the highest resolution of conventional spectroscopy, some of the fine structure of the Balmer β line of deuterium appears. (c) An early saturated absorption spectrum shows clearly improved resolution. (The 1S–2S two-photon resonance is included for reference.) (d) Two recent polarization spectroscopy traces at the Balmer α line of hydrogen (from Refs. [3, 34]).

The history of this improvement can be seen from Fig. 3.22. In emission and absorption, the blue Balmer line of the heavy hydrogen isotope deuterium has a Doppler width that nearly masks the fine structure splitting. With an early pulsed laser saturation spectroscopy system, Hänsch and his co-workers were able to resolve the individual fine structure components [3]. A spectrum of the red Balmer line at this resolution led to an order-of-magnitude improvement in the Rydberg constant. The saturated absorption spectrum shown in Fig. 3.22 was used as a frequency reference for the Doppler-free two-photon resonance in a measurement of the hydrogen ground-state Lamb shift. The Doppler-free two-photon resonance is also indicated.

With a cw laser and a polarization spectroscopy technique, the sensitivity and resolution were improved by another order of magnitude as shown at the bottom of the figure. From this data, a Rydberg constant of $R_\infty =$

109,737.316,76 (32) cm^{-1} was obtained. The chief uncertainty results from an uncertainty in the location of the iodine hyperfine component used as a reference point [34].

The splitting between some of the fine structure components in Fig. 3.22 is due to a quantum electrodynamical effect—the Lamb shift. While this shift can be measured using microwave techniques, the hydrogen saturation spectra were the first to reveal it optically. The isotope shift between hydrogen and deuterium absorption lines can be measured precisely enough with this technique to reduce the uncertainty in the ratio of nucleon to electron masses. Other small shifts predicted quantum mechanically lie just beyond the present limits of precision.

The sensitivity of these spectra to collisional effects such as pressure broadening and shift as well as perturbations due to electrical and magnetic fields implies that saturation spectroscopy can be profitably employed for plasma diagnostics. Peculiar effects are seen for helium buffer gas pressures below 0.1 Torr. In this range, the linewidth of the $2S_{1/2}$–$3P_{1/2}$ transition *narrows* with increasing pressure and the center frequency shifts toward the blue before reversing. These effects can be explained as the result of collisional mixing of hyperfine structure levels somewhat analogous to the "motional narrowing" in NMR and microwave spectroscopy [35].

The other atoms studied extensively either are useful as laser media or have lines which fall conveniently within the gain bands of tunable lasers. A long-term program to reference optical frequency sources to the cesium frequency standard has led to precise values for the Lamb dip frequencies in a number of species. At present, the highest measured atom Lamb dip frequency is that of the ^{20}Ne 1.15-μm laser line at 260.103264 (30) THz. The second harmonic of this laser overlaps a number of I_2 hyperfine components. The absolute frequency of the O$^-$ component of the P(62) 17-1 line of $^{127}I_2$ is presently measured as 520,206,837 $\pm$ 60 MHz [36]. In principle, one can define the speed of light in terms of the vacuum wavelength of this transition and thereby redefine the meter in terms of a frequency measurement.

3.8 MULTIPHOTON AND DOUBLE-RESONANCE SATURATION TECHNIQUES

There is no fundamental reason why saturation spectroscopy should be limited to the single photon transitions discussed in Sections 3.1 and 3.7. The theory of multiquantum saturation spectroscopy follows directly from the theory in Section 3.1 with the substitution of the effective dipole operators of Section 2.2. Moreover, the crossover resonances discussed in Section 3.2 can be thought of as a double-resonance effect and exploited spectroscopically with photons of different frequency.

There are, however, two cases. When one of the photons has a wavelength larger than the sample, one need not be concerned with propagation effects. This radio-frequency photon then acts to shift the frequency of the effective Hamiltonian, making more transitions accessible for study with fixed frequency lasers. The integrals over the Doppler widths in Section 3.1 are unchanged. These phenomena were widely exploited to access states with the same parity as the initial state and measure atomic and molecular term values [37]. Some of the intrinsic precision of radio-frequency spectroscopy is preserved in these optical (or IR)—rf saturation techniques. The transition probabilities need not be small as the frequencies can be chosen to put an intermediate state near resonance.

The other case is typified by the Lamb dips observed by Owyoung in stimulated Raman gain spectroscopy with four counterpropagating beams.

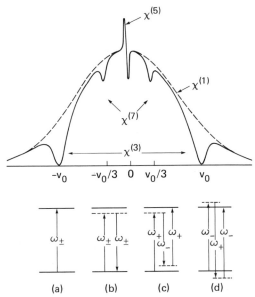

Fig. 3.23 Saturation resonances expected when CH_3F inside a single-frequency laser cavity is probed by a tunable weak unidirectional beam. The two counterpropagating pump waves interact with velocity groups at $\pm v_0$, and the wide troughs centered there and labeled $\chi^{(3)}$ are the power-broadened Lamb dips. (a) The relevant level diagram. (b) The dispersion-shaped "Rayleigh resonance" at zero axial velocity results from the "coherence effects" of Eq. (3.1.18). Formally, this resonance is fifth order in the incident amplitudes and is labeled $\chi^{(5)}$. (c, d) The two velocity-tuned resonances at $\pm v_0/3$ are pumped by three-photon processes. These features result from coherence effects seventh order in the incident amplitudes and are thus labeled $\chi^{(7)}$. In the usual saturation spectroscopy schemes, all of these resonances overlap (from Ref. [39]).

3.8 Multiphoton and Double-Resonance Saturation Techniques

In such an experiment, the momentum of each photon must be considered, and the effective Hamiltonian operator varies with position as $\exp(\pm i(\mathbf{k}_1 - \mathbf{k}_2) \cdot \mathbf{r})$. The formal analysis is more complex, but as in the one-photon case, a dip occurs when the Hamiltonian acts resonantly on molecules with zero axial velocity. The width of the dip reflects the homogeneous linewidth of the net multiphoton transition [38].

Oka has employed this sort of multifrequency technique to separate the features due to the coherence effects of Eq. (3.1.18) from the ordinary Lamb dip [39]. In this experiment, a cell of CH_3F inside the cavity of one laser was pumped by two strong counterpropagating waves and probed by a weaker unidirectional beam from a separate laser. The expected spectrum and explanation is shown in Fig. 3.23. The wide dips marked $\chi^{(3)}$ result from the probe laser interacting with the Bennet holes produced by the pump waves. The dispersion shaped feature labeled $\chi^{(5)}$ results from the "Rayleigh resonance" part of the coherence effect. The three-photon transitions in Figs. 3.22c and 3.22d result in the features labeled $\chi^{(7)}$. They are true multiphoton Lamb dips in this case, but would add a portion of the coherence effect if only one frequency were present. The corresponding experimental results, with the features labeled as above, is shown in Fig. 3.24. Again, the strength of these multiphoton features results from the near-resonant nature of the interaction. The widths are characteristic of the homogeneous widths of the transitions. Under the conditions of these experiments, the Bennet holes were substantially power broadened.

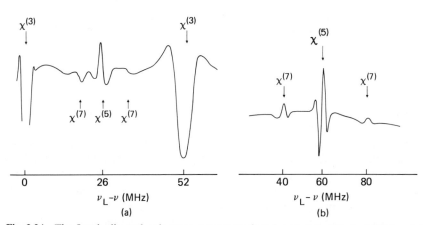

Fig. 3.24 The Lamb dip and coherence resonances found experimentally in CH_3F under the conditions described in Fig. 3.23. Again, the Lamb dips are labeled $\chi^{(3)}$ while the "Rayleigh" and "three-photon pumped multiphoton" resonances are labeled $\chi^{(5)}$ and $\chi^{(7)}$, respectively (Ref. [39]).

3.9 SATURATION TECHNIQUES FOR CONDENSED PHASES

The atoms in a liquid or solid interact strongly with one another and with vibrational modes, but they alter their average position much more slowly than do the atoms of a gas. At low enough temperatures, some vibrational effects vanish. If the other interactions remain constant for a time comparable to the excited state lifetime, the transitions can be inhomogeneously broadened. This broadening results from the random shifts of the energy levels of individual atoms and molecules due to the random local environments in which they sit. These transitions between electronic levels that do not immediately result in phonon emission or require phonon absorption are the optical analogs of the Mossbauer effect. Such transitions are often termed "zero phonon lines" to differentiate them from the broader and stronger "sidebands" due to phonon-assisted electronic transitions. Zero phonon lines are not rare—the ruby R lines are familiar examples—but the recognition that their linewidths are inhomogeneous at low temperatures and that saturation techniques can eliminate most of the width has come recently.

Since there is no Doppler effect to separate the frequencies of the pump and probe laser, saturation experiments in condensed matter require two laser frequencies. In other respects, the theory and practice is quite similar to the Doppler-broadened case. An additional complication, however, results from the fact that the dipole moments for many well-known zero phonon lines are quite weak. The absorption and emission is detectable only because the density obtainable in a solid is much greater than in a gas. While the nonlinear effects scale linearly with the density, they scale as the fourth power of the dipole moment. This fact, and the relatively rapid relaxation rates frequently encountered have tended to limit saturation spectroscopy in condensed matter.

In systems where Bennet holes burned by a pump laser require many milliseconds to recover, saturation spectroscopy experiments can be performed with a single tunable laser. In these systems, the excited levels decay initially to a long-lived "bottleneck" state that can be a low-lying triplet, a state with a different nuclear spin configuration, or some kind of metastable photoproduct. In any case, population in these long-lived states cannot contribute to absorption, and a hole remains in the absorption profile. This hole can be scanned by the same laser (at reduced power) and the change in the absorption detected by monitoring the transmitted beam or the fluorescence from the sample. As in the former case, the width of the hole is twice the homogeneous linewidth, and can be thousands of times narrower than the inhomogeneously broadened absorption profile.

3.9 Saturation Techniques for Condensed Phases

A simple experiment of this type is diagrammed in Fig. 3.25 along with sample data [40]. In the hole-burning stage of the experiment, the frequency of the cw dye laser is held fixed while the shutter opens allowing the full beam intensity to fall on the sample—in this case a crystal of praseodymium-doped lanthanum trifluoride immersed in liquid helium at 2°K. After the shutter closes, an attenuator ($ND > 2$) is automatically inserted in the beam, and the shutter opens with the frequency sweeping rapidly over the hole position. The fluorescence on a coupled transition is plotted as a function of laser frequency.

The insert in Fig. 3.25 clearly shows the resulting hole at zero frequency, the width of which is 2 MHz (FWHM), much less than the 3 GHz of the inhomogeneously broadened line. Also present are satellite holes and "antiholes" due to crossover transitions to other hyperfine levels. The holes, in this case, are quite long-lived as their recovery requires equilibration among nuclear hyperfine states with lifetimes of many minutes. Such long-lived

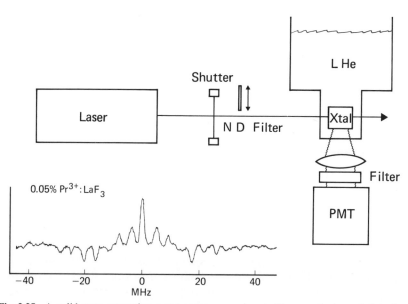

Fig. 3.25 A solid-state saturation spectroscopy experiment. The sample is cooled to low temperature with liquid helium to suppress phonon effects. A hole is first burned in the absorption line using the full laser power with the frequency stabilized. The shutter is then closed and the neutral density filter automatically inserted to reduce the intensity of the probe wave. When the probe frequency is then tuned across that of the hole, a dip appears in the fluorescent sidelight. The inset shows a pattern of holes and "antiholes" resulting from nonequilibrium populations in various hyperfine levels of the ground state produced by hole burning.

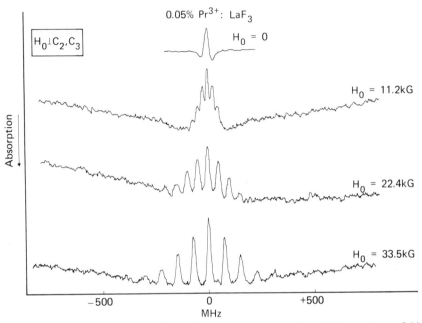

Fig. 3.26 Zeeman splitting on the $^3H_4 \to {}^1D_2$ transition in Pr:LaF$_3$ at 592.5 nm. At zero field the fine structure shown in the inset of Fig. 3.25 is unresolved. When a magnetic field is applied after the initial holes are burned, features due to individual Zeeman levels separate. The inhomogeneous linewidth of this transition is roughly 3 GHz. Vastly larger magnetic fields would be required to observe Zeeman splitting without hole burning (from Ref. [41]).

holes can be manipulated by applying electric or magnetic fields or mechanical stress. The lower traces in Fig. 3.26 show the splitting that results from a magnetic field applied after the hole is burned. Linear and quadratic Zeeman effects can be measured for the excited state [41].

Very similar results can be obtained from a polarization spectroscopy technique in which a polarized pump beam is used to bleach absorption centers with a definite dipole moment orientation, and the resulting changes in birefringence and dichroism probed by a second beam polarized at 45° to the first. The theory is similar to that for isotropic vapors except that the symmetries of the absorbing sites and of the crystal must be taken into account.

Long-lived holes appear rather common at low temperatures. In some photochemical systems, the lifetime appears infinite. Free base porphin in alkane matrices exhibits stable holes at low temperatures. These molecules have two tautomeric forms, the energy levels of which are split by the interaction with the matrix. These two forms are illustrated in Fig. 3.27. From the excited state, a molecule can decay into either of the two forms, leaving

3.9 Saturation Techniques for Condensed Phases

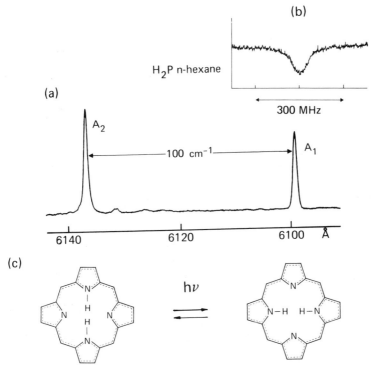

Fig. 3.27 A typical molecule showing photochemical hole burning, its absorption spectrum, and a typical hole. (c) The molecule—free-base porphin—has two forms depending upon the positions of the internal hydrogen atoms. (a) In a Shpolskii matrix composed of microcrystals of n-hexane, the energy levels of these two forms are split by the reduced symmetry of the molecular environment. In the excited state, the hydrogens can reorient, but a potential barrier holds them fixed in the ground state. (b) A typical photochemical hole burned in the A_1 component. Such narrow features require temperatures below $2°K$ (from Ref. [42]).

a narrow hole in the initial state and producing a broad antihole on the absorption band of the other tautomer. Rotation of the two hydrogens is inhibited by an energy barrier; at low temperatures holes persist as long as patience and liquid helium are available [42].

A related technique requiring only one laser is termed laser-induced fluorescence line narrowing. A similar phenomenon in gases has long been known, but in the case of condensed systems, narrow fluorescence can be observed at all angles rather than only in the forward direction. In these experiments, a laser is used to excite centers with resonant frequencies near the laser frequency. After the laser is turned off, these centers decay, with a fluorescence linewidth less than the absorption linewidth. If the fluorescence is observed on a resonance line, the width is essentially the homogeneous width. For other transitions, the differences in the effects of strain on different levels can lead to an inhomogeneous component. In either case, the fluorescence is usually resolved in frequency with a Fabry–Perot interferometer, and occasionally in time as well. Spectral diffusion in these systems is related to energy transfer between ions, and time-resolved fluorescence line-narrowing experiments is an unequivocal method of observing such effects [43].

Yen et al. have eliminated the residual inhomogeneous broadening component with the three-level fluorescence line-narrowing scheme illustrated in Fig. 3.28 [44]. One laser excites the upper level $|a\rangle$ which ordinarily would emit fluorescence to level $|c\rangle$. A second laser beam on the $|a\rangle$, $|c\rangle$ transition depopulates level $|a\rangle$ by simulated emission, leading to a hole in the spectrum of fluorescence sidelight that is resolved by a Fabry–Perot interferometer. The width of the hole in Figs. 3.27b and 3.27c yields the homogeneous linewidth of the $|a\rangle$, $|c\rangle$ transition.

Saturation spectroscopy of short-lived systems requires pulsed lasers with independent pump and probe frequencies. A polarization spectroscopy

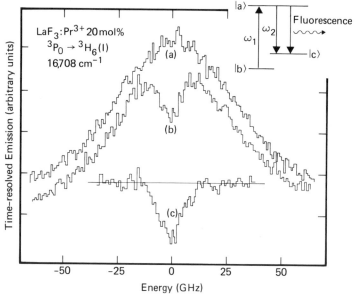

Fig. 3.28 Eliminating residual inhomogeneous broadening in laser-induced fluorescence with hole burning. (a) The normal laser-induced fluorescence line-narrowing spectrum. (b) The result of stimulating the emission from $|a\rangle$ to $|c\rangle$ with a narrow-band laser. Ions tend to decay by stimulated emission rather than by fluorescence which results in a hole in the fluorescence spectrum as resolved by a Fabry–Perot interferometer. Subtracting traces (a) and (b) results in (c) the hole spectrum which reflects the true homogeneous linewidth (from Ref. [44]).

3.9 Saturation Techniques for Condensed Phases

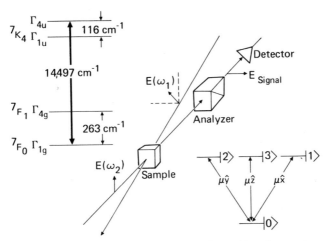

Fig. 3.29 A solid-state polarization spectroscopy experiment. Two frequencies are required for experiments in which the excited state lifetime is shorter than the laser pulses. The pump wave is at ω_1 while the weaker probe has frequency ω_2. The polarization analyzer blocks the probe except when pump and probe interact with the same absorbing centers. The level diagram shown applies to $Sm^{2+}:CaF_2$ in which the upper level is degenerate (from Ref. [45]).

experiment of this sort is illustrated in Fig. 3.29 [45]. When two fields of different frequency simultaneously interact with the sample, there is an additional effect: the two fields can be combined into a single amplitude modulated wave. If the modulation rate is slow compared to the lifetimes of the states involved, the population difference follows the modulation adiabatically. The resulting fluctuation in the complex index of refraction of the medium scatters some of the pump light into the detection channel, giving rise to a signal related to the "Rayleigh resonance" in Eq. (2.4.27). If the modulations are too fast for the populations to follow, the effect disappears. The structure of the resulting peak at zero frequency difference between the pump and probe reflects the longitudinal decay rates or lifetimes of the states involved. Sargent, Fayer, Yajima, and others have employed these "moving grating effects" to measure picosecond and shorter lifetimes with nanosecond pulse-length lasers [46–48].

The expected and observed line shapes in the experiment in Fig. 3.29 at 35°K are shown in Fig. 3.30. The narrow spike is the Rayleigh resonance or moving grating effect, while the broad peak reflects the homogeneous linewidth of the transition. At lower temperatures, the homogeneous linewidth becomes a factor of 6 smaller, and the narrow spike disappears from the polarization spectroscopy signal. The theory of these effects can become quite involved especially in systems with orientationally degenerate levels.

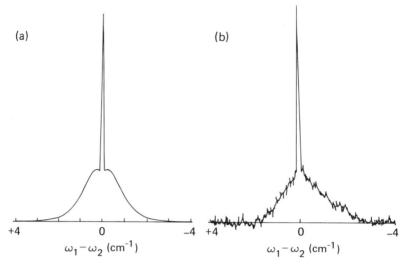

Fig. 3.30 A comparison of the (a) theoretical and (b) experimental polarization spectroscopy lineshapes in $Sm^{2+}:CaF_2$, 0.003%, 35°K. The width of the narrow central peak reflects longitudinal relaxation rates, while the full width of the broader peak is four times the transverse decay rate (from Ref. [45]).

3.10 APPLICATIONS OF SATURATION TECHNIQUES

Outside the laboratory, saturation techniques have potential applications only slightly related to high-resolution spectroscopy and well-defined wavelength and frequency standards. The most readily understood is in the area of computer memory and data storage. At present, there are optical techniques of storing information similar to the video-disk technology that can store one bit of information per square wavelength of surface area. Basically, a tightly focused laser beam is used to burn a hole in a thin film at a location that corresponds to a "bit." A second focused laser can detect that hole and read the "bit."

If the surface were coated with a medium having an inhomogeneously broadened absorption line, and if the bit-writing laser were tunable, it could write many bits at the same spatial location. Each bit should correspond to a Bennet hole at a particular frequency within the inhomogeneously broadened line. The reading laser would scan the line, reading all of the bits. In principle, the storage density of such a system can be 10,000 times better than the best video-disk technology. Materials supporting long-lived Bennet holes are presently available; the need for cryogenic temperatures may not be an insuperable barrier.

3.10 Applications of Saturation Techniques

Optical bistability is a second application related to data processing [49]. An interferometer filled with a saturable medium would have poor transmission because of the absorption and dispersion of the medium. If, however, the intensity inside the interferometer is high enough to saturate the medium, the transmission dramatically increases. Once this saturation condition is reached, the resonant build-up of the intensity inside the interferometer is sufficient to maintain saturation even if the incident intensity is reduced. Both the high-transmission and low-transmission states are stable at the same incident intensity. The result is a bistable device similar to an optical toggle switch, or flip flop. Optical transistor action can also be obtained by employing a weak beam to trigger saturation in an interferometer biased near the threshold for turning on. Such devices can be used to process data in an optical computer, even though at present the speed and power consumption do not compare favorably with electronic devices.

Wave-front conjugation by degenerate four-wave mixing is an almost magical application of the Rayleigh resonance phenomenon [50]. In degenerate four-wave mixing, two pump beams propagate in opposite directions through a saturable medium. An object beam carrying information is also incident. It forms an interference pattern with one of the pump waves. At the bright fringes of the pattern, the absorption is saturated and the complex index of the refraction of the medium is perturbed. At the dark fringes, there is no saturation. The other pump wave can diffract from the index perturbations due to the fringe pattern, and the diffracted wave propagates out of the medium as the phase conjugate of the object wave front. As a function of time, the phase conjugate wave propagates as the object wave would propagate *if time were reversed*. A properly located beam splitter can intercept part of the conjugate wave and direct it elsewhere. If the object wave had propagated through a pattern of apertures, the image wave would project an identical pattern at a corresponding location.

The remarkable fact is that this image remains even if the object beam is scrambled by propagating through an aberrating medium. The conjugate wave propagates back through the aberrating medium along exactly the same ray paths followed by the object wave. The scrambling is thus undone. Such adaptive optical devices are useful whenever high power laser beams must be brought to a tight focus (as in laser fusion), transmitted through the atmosphere, or used to produce a pattern. The images produced are of higher quality than can be obtained by conventional optics or holography. An image projector apparatus of this sort employing ruby as a saturable material is illustrated in Fig. 3.31. Also shown is a typical projected image [51]. Truly practical applications presently await invention.

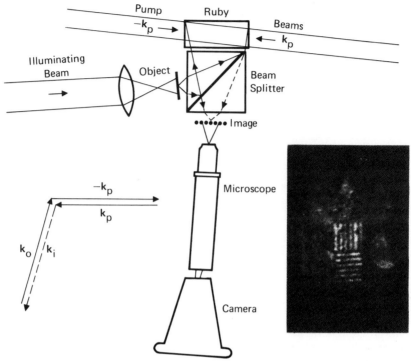

Fig. 3.31 Conjugate wave-front generation by degenerate four-wave mixing in ruby used to project images. The object and pump beams interfere in the ruby crystal to produce regions of high and low intensity. Because the high-intensity regions are more saturated than the low-intensity regions, index-of-refraction perturbations result which scatter the pump beams into an image wave that reproduces the object wave front with very high fidelity. The horizontal bars in the inset are 5 μm wide (from Ref. [51]).

REFERENCES

1. V. S. Letokhov and V. P. Chebotayev, "Nonlinear Laser Spectroscopy" (Springer Series in Optical Sciences 4). Springer-Verlag, Berlin, 1977.
2. V. P. Chebotayev and V. S. Letokhov, Nonlinear narrow optical resonances induced by laser radiation, *in* "Progress in Quantum Electronics" (J. H. Sanders and S. Stenholm eds.), Vol. 4, pp. 111–206. Pergamon, Oxford, 1977.
3. T. W. Hänsch, Nonlinear high resolution spectroscopy of atoms and molecules, *in* "Nonlinear Optics" (N. Bloembergen, ed.) (Proc. Int. School Phys., Enrico Fermi Course 64), pp. 17–86. North-Holland Publ., Amsterdam, 1977.
4. V. P. Chebotayev, Three level laser spectroscopy, *in* "High Resolution Laser Spectroscopy" (K. Shimoda, ed.) (Topics in Applied Physics 13), pp. 207–251. Springer-Verlag, Berlin, 1976.

References

5. W. Demtroder, "Laser Spectroscopy," (Chemical Physics 5). Springer-Verlag, Berlin, 1981.
6. T. W. Hänsch and P. Toschek, Theory of a three level gas laser amplifier, *Z. Physik* **236**, 213–244 (1970).
7. V. S. Letokhov, Saturation spectroscopy, *in* "High Resolution Laser Spectroscopy" (K. Shimoda, ed.) (Topics in Applied Physics 13), pp. 95–173. Springer-Verlag, Berlin, 1976; J. L. Hall and J. A. Magyar, High resolution saturation absorption studies in methane and some methylhalides, *ibid.*, pp. 174–199.
8. B. J. Feldman and M. S. Feld, *Phys. Rev. A* **1**, 1375 (1970); **5**, 899 (1972).
9. F. Y. Wu, S. Ezekial, M. Ducloy, and B. R. Mollow, *Phys. Rev. Lett.* **38**, 1077–1080 (1977).
10. S. H. Autler and C. H. Townes, *Phys. Rev.* **100**, 703 (1955).
11. S. Haroche and F. Hartman, *Phys. Rev. A* **6**, 1280–1300 (1972).
12. E. V. Baklanov and V. P. Chebotayev, *Soviet Phys.—JETP* **60**, 551 (1971); **61**, 922 (1971).
13. V. P. Chebotayev, Coherence in high resolution spectroscopy, *in* "Coherent Nonlinear Optics" (M. S. Feld and V. S. Letokhov, eds.) (Topics in Current Physics 21), pp. 59–109. Springer-Verlag, Berlin, 1980.
14. C. Wieman and T. W. Hänsch, *Phys. Rev. Lett.* **34**, 1120 (1976).
15. R. Teets, R. Feinberg, T. W. Hänsch, and A. L. Schawlow, *Phys. Rev. Lett.* **37**, 683 (1976). Also N. W. Carlson, A. J. Taylor, K. M. Jones, and A. L. Schawlow, *Phys. Rev. A* **24**, 822–834 (1981).
16. W. Gawlik and G. W. Series, Forward scattering and polarization spectroscopy, *in* "Laser Spectroscopy IV" (H. Walther and K. W. Rothe, eds.) (Springer Series in Optical Sciences 21), Vol. 4, pp. 210–222, Springer-Verlag, Berlin, 1979.
17. C. J. Mullin, J. M. Keller, C. L. Hammer, and R. H. Good, Jr., *Ann. Phys. (NY)* **37**, 55 (1966).
18. S. Saikan, *J. Opt. Soc. Amer.* **68**, 1184 (1978).
19. M. Sargent III, *Phys. Rev. A* **14**, 524–527 (1976).
20. B. Wellegehausen, Optically pumped cw dimer lasers, *IEEE J. Quant. Electron.* **QE-15**, 1108–1130 (1979).
21. A. Szoke and A. Javan, *Phys. Rev. Lett.* **10**, 521 (1963); R. A. McFarlane, W. R. Bennet, Jr., and W. E. Lamb, Jr., *Appl. Phys. Lett.* **2**, 189 (1963).
22. C. Freed and A. Javan, *Appl. Phys. Lett.* **17**, 53 (1970).
23. P. H. Lee and M. L. Skolnick, *Appl. Phys. Lett.* **10**, 303 (1967).
24. T. Hänsch, M. D. Levenson, and A. L. Schawlow, *Phys. Rev. Lett.* **26**, 946 (1971).
25. F. K. Kowalski, W. T. Hill, and A. L. Schawlow, *Opt. Lett.* **2**, 112–114 (1978); R. Schieder, Interferometric nonlinear spectroscopy, *Opt. Commun.* **24**, 113–116 (1978).
26. J. C. Keller and C. Delsart, *Opt. Commun.* **20**, 147 (1977).
27. N. F. Ramsey, *Phys. Rev.* **78**, 695 (1950).
28. J. C. Bergquist, S. A. Lee, and J. L. Hall, *Phys. Rev. Lett.* **38**, 159 (1977).
29. R. L. Barger, J. C. Bergquist, T. C. English, and D. C. Glaze, Resolution of photon recoil structure of the 6573 Å calcium line in an atomic beam with optical Ramsey fringes, *Appl. Phys. Lett.* **34**, 190–191 (1979).
30. C. J. Borde, J. C. Hall, C. J. Kunasz, and D. G. Hummer, *Phys. Rev.* **14**, 236–263 (1976); J. L. Hall and C. J. Borde, *Appl. Phys. Lett.* **29**, 788–790 (1976).
31. R. L. Barger, Influence of second order Doppler effect on optical Ramsey fringe, *Opt. Lett.* **6**, 145–148 (1981).
32. M. D. Levenson and A. L. Schawlow, *Phys. Rev. A* **6**, 946 (1972).
33. G. Camy, B. Decomps, J. L. Gardissat, and C. J. Borde, *Metrologia* **13**, 145–148 (1977).
34. J. E. M. Goldsmith, E. W. Weber, and T. W. Hänsch, *Phys. Rev. Lett.* **41**, 1525 (1978).

35. E. W. Weber and J. E. M. Goldsmith, *Phys. Lett.* **70A**, 95 (1979).
36. K. M. Baird, K. M. Evenson, G. R. Hanes, D. A. Jennings, and F. R. Petersen, Extension of absolute frequency measurements to the visible: Frequencies of ten hyperfine components of iodine, *Opt. Lett.* **4**, 263–264 (1979).
37. F. Shimizu, *Phys. Rev. A* **10**, 950–959 (1974), is the basic theoretical reference; S. M. Freund and T. Oka, *Appl. Phys. Lett.* **21**, 60–62 (1972); S. M. Freund and T. Oka, *Phys. Rev. A* **13**, 2176–2190 (1976); E. Arimondo and P. Glorieux, *ibid.* **19**, 1067–1083 (1979) report typical programs. Many other papers have been published by Oka and the Herzberg Institute Group as well as by Shimoda, Weber, and others.
38. A. Owyoung and P. Esherick, Sub-Doppler Raman saturation spectroscopy, *Opt. Lett.* **5**, 421–423 (1980).
39. J. Reid and T. Oka, *Phys. Rev. Lett.* **38**, 67–70 (1977).
40. R. M. Shelby and R. M. Macfarlane, Measurement of the pseudo-Stark effect in $Pr^{3+}LaF_3$ using population hole-burning and optical free-induction decay, *Opt. Commun.* **27**, 399–402 (1978).
41. R. M. Macfarlane and R. M. Shelby, Measurement of nuclear and electronic Zeeman effects using optical hole-burning spectroscopy, *Opt. Lett.* **6**, 96–98 (1981).
42. S. Völker, R. M. Macfarlane, and J. H. van der Waals, *Chem. Phys. Lett.* **53**, 8 (1978).
43. W. M. Yen and P. M. Seltzer, "Laser Spectroscopy of Solids" (Topics in Applied Physics 49). Springer-Verlag, Berlin, 1981, and references therein.
44. W. M. Yen, private communication.
45. J. H. Lee, J. J. Song, M. A. F. Scarparo, and M. D. Levenson, *Opt. Lett.* **5**, 196–198 (1980).
46. M. Sargent III, P. E. Toschek, and H. G. Danielmeyer, Unidirectional saturation spectroscopy I, *Appl. Phys.* **11**, 55–62 (1976); M. Sargent III and P. E. Toschek, Unidirectional saturation spectroscopy II, *ibid.* 107–120 (1976).
47. T. Yajima, *Opt. Commun.* **14**, 378–382 (1975).
48. J. R. Salcedo, A. C. Siegman, D. D. Dlott, and M. D. Fayer, *Phys. Rev. Lett.* **41**, 131 (1978).
49. Special Issue on optical bistability, *IEEE J. Quant. Electon.* **QE-17**, 300–386 (1981), and references therein.
50. A. Yariv, *IEEE J. Quant. Electon.* **QE-14**, 650–660 (1978); **QE-15**, 524 (1979), and references therein.
51. M. D. Levenson, *Opt. Lett.* **5**, 182–184 (1980).
52. R. G. Brewer, M. J. Kelley, and A. Javan, *Phys. Rev. Lett.* **23**, 559–563 (1969).
53. W. Gawlik, J. Kowalski, F. Trager, and M. Vollmer, *Phys. Rev. Lett.* **48**, 871–873 (1982). *Note Added in Proof*: Some of these narrow resonances have now been shown to have a width *less* than T_2^{-1} and have been attributed to a light-shift-induced Hanle effect.

Chapter 4
COHERENT RAMAN SPECTROSCOPY

INTRODUCTION

While Raman scattering was introduced in Section 1.1 as one of the classical tools of linear spectroscopy, the underlying process clearly involves a nonlinear coupling between matter and electromagnetic radiation. In the conventional, old-fashioned ordinary Raman scattering (COORS) process shown in Fig. 4.1a, an incident photon of energy $\hbar\omega_1$ is destroyed in the medium while at the same time a scattered photon of energy $\hbar(\omega_1 - \Omega_R)$ is emitted. The difference in energy between the two appears as a quantum of excitation of the Raman mode with energy $\hbar\Omega_R$. Stokes scattering is shown in Fig. 4.1a—the emitted energy is less than the incident energy, and thus the internal energy of the scattering medium is increased. Anti-Stokes processes extract energy from excited Raman modes by emitting quanta with energy greater than the incident quanta.

The important point to recall is that two quanta are involved in each transition, and that the process is coherent, properly described by a Fermi golden-rule rate linear in the incident intensity and proportional to a product of dipole matrix elements. As Placzek first pointed out, Raman scattering can be explained as spontaneous emission due to the field-dependent dipole moment defined in Eq. (2.7.17) [1]. The formalism of Section 2.7 can be employed to calculate the scattering cross section in (1.1.14). Spontaneous scattering differs from other two-photon processes only in that it is "stimulated" by the vacuum fluctuations in the electromagnetic field. The observed effects scale linearly with intensity and are observable even with relatively weak light sources.

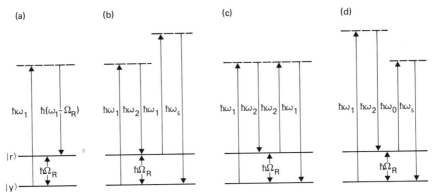

Fig. 4.1 Level diagrams in Raman spectroscopy. These illustrations only compare energy level splittings to photon energies and should not be interpreted as defining a time-ordered processes or population transfers. (a) The spontaneous scattering mechanism operative in conventional old-fashioned ordinary Raman spectroscopy (COORS). (b) The process in coherent anti-stokes Raman spectroscopy (CARS). The output frequency is denoted ω_s and by convention $\omega_1 > \omega_2$. Coherent Stokes Raman spectroscopy (CSRS) would have the same diagram with $\omega_1 < \omega_2$. (c) The process operative in stimulated Raman gain (SRG) and the inverse Raman effect (TIRE) as well as the Raman-induced Kerr effect (TRIKE). (d) The process of four-wave mixing with three input frequencies. All resonances are at $\omega_1 - \omega_2 = \Omega_R$.

Much of the nomenclature of coherent Raman spectroscopy has been presented in Section 1.3.2 and in Fig. 1.10 while the detailed theory appears in Chapter 2, particularly Eqs. (2.4.21) and Sections 2.5 and 2.7–2.9. There are also a number of detailed review articles in the literature [2–8]. What follows here is a simplified summary intended to provide some intuitive understanding of the effects occurring in the most common case, the case in which resonant effects are absent.

4.1 DRIVING AND DETECTING A RAMAN MODE

If the energies of the incident photons are all much less of the energies of the dipole-allowed transitions in a system, all of the denominators in the expressions for the coupling Hamiltonian and dipole operator in (2.7.12) to (2.7.17) become equal. In this Born–Oppenheimer approximation, one can parametrize the strength of the Raman interaction with the tensor element

$$\alpha^R_{\alpha\beta} = \frac{\langle g|\tilde{\mu}_\alpha|n\rangle\langle n|\tilde{\mu}_\beta|r\rangle + \langle g|\tilde{\mu}_\beta|n\rangle\langle n|\tilde{\mu}_\alpha|r\rangle}{\hbar\Omega_{ng}}, \qquad (4.1.1)$$

where the superscript R denotes the Rth Raman mode, that is, the mode

4.1 Driving and Detecting a Raman Mode

corresponding to the transition between the lower level $|g\rangle$ and the excited state $|r\rangle$ [2, 4, 9].[1]

In coherent Raman spectroscopy, there are always at least two incident frequencies. Assuming that the difference between two of them approaches the frequency of the mode R (i.e., $|\omega_1 - \omega_2| \approx \Omega_R = \hbar^{-1}(E_r - E_g)$), the coupling Hamiltonian and dipole operator become

$$\tilde{\mathcal{H}}_{\omega_1 - \omega_2} = -\tfrac{1}{4}\alpha^R_{\gamma\delta}E_\gamma(\omega_1)E^*_\delta(\omega_2),$$

$$\tilde{\mu}_\alpha(t) = -\tfrac{1}{2}\alpha^R_{\alpha\beta}\sum_{\omega_0}\{E_\beta(\omega_0)e^{-i\omega_0 t} + E^*_\beta(\omega_0)e^{i\omega_0 t}\}, \qquad (4.1.2)$$

and the equation of motion for the off-diagonal matrix element [i.e., (2.4.5c)], becomes

$$\dot{\rho}_{gr} = (-i/16\hbar)(\rho_{rr} - \rho_{gg})\alpha^R_{\gamma\delta}E_\gamma(\omega_1)E^*_\delta(\omega_2)e^{-i(\omega_1 - \omega_2)t} - i\Omega_R\rho_{gr} - \rho_{gr}/T_2. \qquad (4.1.3)$$

Defining the real-valued Raman coordinate by $Q = \rho_{rg} + \rho_{gr}$, one can derive an equation of the form in (1.3.5) by assuming that the steady-state population difference between levels $|g\rangle$ and $|r\rangle$ remains unperturbed by this interaction. Writing the relevant population difference as w^e, the equation of motion for Q_R assumes the familiar form

$$\ddot{Q} + 2\dot{Q}/T_2 + \Omega_Q^2 Q = (w^e\Omega_{rg}/2\hbar)\alpha^R_{\gamma\delta}E_\gamma E^*_\delta e^{-i(\omega_1 - \omega_2)t}, \qquad (4.1.4)$$

where $\Omega_Q^2 = \Omega_{rg}^2 - T_2^{-2}$. The coordinate Q can be interpreted in various ways, including the displacement in a mass-on-spring model of molecular vibration. Taking the steady-state solution

$$Q(t) = Q(\omega_1 - \omega_2)e^{-i(\omega_1 - \omega_2)t} + \text{cc} \qquad (4.1.5)$$

plus similar terms with other frequency differences, we have

$$Q(\omega_1 - \omega_2) = \frac{w^e\Omega_{rg}\alpha^R_{\gamma\delta}E_\gamma(\omega_1)E^*_\delta(\omega_2)}{2\hbar[\Omega_Q^2 - (\omega_1 - \omega_2)^2 - 2iT_2^{-1}(\omega_1 - \omega_2)]}. \qquad (4.1.6)$$

In analogy with (1.3.6), the oscillating Raman coordinate drives a radiating polarization

$$P_\alpha^Q(t) = N\tilde{\mu}_\alpha(t)Q(t) = \sum_{\omega_0}\{\tfrac{1}{2}P_\alpha^Q(\omega_0 + \omega_1 - \omega_2)e^{-i(\omega_0 + \omega_1 - \omega_2)t}$$

$$+ \tfrac{1}{2}P_\alpha^Q(\omega_0 + \omega_2 - \omega_1)e^{-i(\omega_0 + \omega_2 - \omega_1)t} + \text{cc}\}. \qquad (4.1.7)$$

---

[1] Again in condensed matter, the dipole moment operator must be corrected for local field effects, $\tilde{\mu} = (n^2 + 2)e\tilde{r}/3$.

Each Fourier component of this polarization radiates a different coherent Raman signal, but all are related to one another. Several such signals are illustrated in Fig. 1.10.

The amplitudes of the Fourier components are most conveniently expressed in terms of the third-order nonlinear susceptibility tensor,

$$P_\alpha^Q(\omega_s) = D\chi_{\alpha\beta\gamma\delta}^Q(-\omega_s, \omega_0, \omega_1, -\omega_2)E_\beta(\omega_0)E_\gamma(\omega_1)E_\delta^*(\omega_2), \quad (4.1.8)$$

where $\omega_0 + \omega_1 - \omega_2 - \omega_s = 0$. The combinational factor D is the number of distinguishable combinations of the incident field amplitudes. When all the fields are distinguishable in frequency or polarization, $D = 6$; but the combinatorial factor is reduced to 3 when a single field occurs twice in the interaction [2, 9].

The Raman resonant nonlinear susceptibility is complex, implying that the polarization P^Q can have a phase shift with respect to the driving fields which depends upon the detuning from resonance. The explicit form for the contribution to the nonlinear susceptibility resonant at $|\omega_1 - \omega_2| = \Omega_Q$ is

$$\chi_{\alpha\beta\gamma\delta}^Q(-\omega_s, \omega_0, \omega_1, -\omega_2) = \frac{\mathcal{N} w^e \Omega_{rg} \alpha_{\alpha\beta}^R \alpha_{\gamma\delta}^R}{12\hbar[\Omega_Q^2 - (\omega_1 - \omega_2)^2 - 2i(\omega_1 - \omega_2)/T_2]}. \quad (4.1.9)$$

Similar contributions occur when $|\omega_0 - \omega_2| = \Omega_Q$, etc. In the case of CARS and CSRS, two incident frequencies are equal and the two contributions always occur in pairs symmetric with respect to the polarization subscripts β and γ,

$$\chi_{\alpha\beta\gamma\delta}^Q(-\omega_s, \omega_1, \omega_1, -\omega_2) = \frac{\mathcal{N} w^e \Omega_{rg}(\alpha_{\alpha\beta}^R \alpha_{\gamma\delta}^R + \alpha_{\alpha\gamma}^R \alpha_{\beta\delta}^R)}{12\hbar[\Omega_Q^2 - (\omega_1 - \omega_2)^2 - 2i(\omega_1 - \omega_2)/T_2]}. \quad (4.1.10)$$

Equations (4.1.9) and (4.1.10) describe all of the coherent Raman spectroscopy resonances in transparent media, and show the proper transformation symmetries under interchange of frequency subscripts, etc.

The total nonlinear susceptibility tensor includes a nonresonant contribution due to virtual transitions as well as a sum of Raman resonant terms,

$$\chi_{\alpha\beta\gamma\delta}^{(3)} = \chi^{NR} + \chi^R = \chi^{NR} + \sum_Q \chi^Q. \quad (4.1.11)$$

Generally, the nonresonant term can be treated as independent of the frequency difference $\omega_1 - \omega_2$. When the Born–Oppenheimer approximation applies, χ^{NR} is a real constant [4].

This formalism describes a variety of coherent Raman processes each producing a distinct signal frequency and each related to a definite set of frequency arguments for the nonlinear susceptibility. The energies of some

4.1 Driving and Detecting a Raman Mode

of the input and output photons are compared to material energy levels in Figs. 4.1 and 1.9. Some of the spectroscopic tools based upon these effects are represented in Fig. 1.10. Each case corresponds to a process of the form shown in Fig. 4.2; the effects driving the resonant vibration are always the same, but the field scattered by the oscillating Raman mode according to (4.1.6) defines the coherent Raman technique [2].

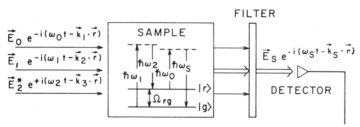

Fig. 4.2 Schematic of a general coherent Raman process. The input waves mix together in the sample to produce a signal amplitude E_s that is separated from the inputs and detected. Two of the input frequencies are often equal: in CARS $\omega_0 = \omega_1$ while in TRIKE, TIRE, and SRG $\omega_0 = -\omega_1$ (from Ref. [2]).

The history of coherent Raman spectroscopy began with the discovery of stimulated Raman oscillation by Woodbury and Ng [10]. While useful for generating intense beams at wavelength shifted from the incident frequency by a multiple of the Raman frequency, this effect has had few direct applications in spectroscopy. Later "stimulated Raman gain" techniques restricted themselves to the low-gain regime well below the threshold for oscillation [6]. If $\omega_1 > \omega_2$, $\chi^{(3)}(\omega_2, -\omega_1, \omega_1, -\omega_2)$ describes the nonlinearity. The stimulated Raman effect that produces oscillation at Stokes-shifted frequencies also produces loss at the corresponding anti-Stokes frequency. This loss, first demonstrated by Jones and Stoicheff using a narrow-band pump and a broad-band probe, has been used extensively in spectroscopy and is now generally termed "the inverse Raman effect" (TIRE) [11]. If $\omega_1 > \omega_2$, the correct nonlinear susceptibility is $\chi^{(3)}(\omega_1, -\omega_2, \omega_2, -\omega_1)$.

The three-wave mixing process in which new frequency components are generated were first demonstrated by Yajima and by Maker and Terhune [9]. Stansfield et al. employed tunable dye lasers in an early four-wave mixing experiment. Ultimately, these techniques were renamed "coherent anti-Stokes Raman spectroscopy" (CARS), and "four-wave CARS," by Begley et al. and Compaan et al. The relevant nonlinear susceptibilities are $\chi^{(3)}(\omega_s, \omega_1, \omega_1, -\omega_2)$ and $\chi^{(3)}(\omega_s, \omega_0, \omega_1, -\omega_2)$, respectively. Still later, Hellwarth proposed and Heiman et al. demonstrated a truly background-free spectroscopy based upon "the Raman induced Kerr effect" (TRIKE) [12].

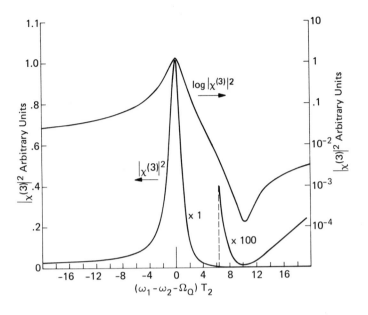

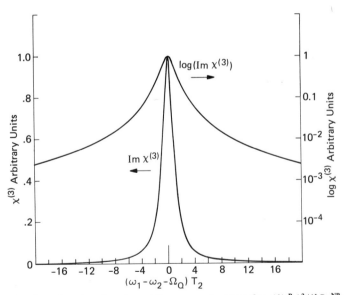

Fig. 4.3 Line-shape functions in coherent Raman spectroscopy for $\mathcal{N}(\alpha_{\text{eff}}^{R})^2/4\hbar D\chi_{\text{eff}}^{\text{NR}} = 10/T_2$ on linear and logarithmic scales. A single isolated Lorentzian resonance has been assumed. (a) In CARS, CSRS, and 4 WM an interference between resonant and nonresonant susceptibilities produces the distinct minimum. (b) The COORS, SRG, TIRE, and RIKES (circular polarization) line shape is symmetric (from Ref. [2]).

The nonlinear susceptibility which describes these effects have the same frequency arguments as SRG and TIRE. The history of this field has been reviewed extensively, most notably by Bloembergen [13], by Shen [14], and by Levenson and Song [2].

All of the observed effects are properly described by a nonlinear susceptibility of the form in (4.1.11), but some of the detection schemes respond to $|\chi^{(3)}|^2$ while others detect the real or imaginary parts of $\chi^{(3)}$ separately. Thus, the line-shape functions are not quite identical, even if the underlying physics is. The line-shape functions produced by an isolated Lorentzian Raman resonance in CARS, stimulated Raman gain spectroscopy, and RIKES are shown in Fig. 4.3. The details of these line shapes will be discussed more extensively in later sections.

Because of the presence of the nonresonant contribution to many CRS signals, and because these techniques are often used to probe condensed phases, greater importance has been attributed to the symmetry properties of $\chi^{(3)}$ in coherent Raman spectroscopy than in other areas. To cope with the complexity of a fourth-rank tensor with three independent frequency arguments, a standard convention has been adopted for the arguments and subscripts of $\chi^{(3)}_{\alpha\beta\gamma\delta}(-\omega_s, \omega_0, \omega_1, -\omega_2)$. In the usual Maker–Terhune convention, the arguments and subscripts can be freely permuted so long as the pairing of argument and subscript is respected [9]. In the CRS convention, the order of the arguments is fixed [2]. In this convention, the arguments can be suppressed and different susceptibility elements can be indicated by merely altering the subscripts.

4.2 SYMMETRY CONSIDERATIONS

Chi-three is a fourth-rank tensor with up to 81 nonvanishing elements. The symmetry of the nonlinear medium reduces the number of independent nonvanishing elements. In an isotropic material, there are four nonvanishing elements

$$\chi^{(3)}_{1111}, \quad \chi^{(3)}_{1122}, \quad \chi^{(3)}_{1212}, \quad \text{and} \quad \chi^{(3)}_{1221}$$

which must fulfill the relationship

$$\chi^{(3)}_{1111} = \chi^{(3)}_{1122} + \chi^{(3)}_{1212} + \chi^{(3)}_{1221}. \tag{4.2.1}$$

The numerical subscripts refer to the axes of an arbitrary Cartesian coordinate system. For the 32 point-group symmetries, the relationships among the nonvanishing elements of $\chi^{(3)}$ appear in Table 4.1, as tabulated by Butcher.

When the optical frequencies are much less than the frequencies of electronic transitions, the frequency arguments of the background (non-Raman

TABLE 4.1 *Form of Third-Order Nonlinear Susceptibility Tensor $\chi^{(3)}_{\alpha\beta\gamma\delta}(-\omega_s, \omega_0, \omega_1, -\omega_2)$ for the 32 Crystal Classes and Isotropic Media*[a]

Triclinic
For both classes 1 and $\bar{1}$, there are 81 independent nonzero elements.

Monoclinic
For all three classes 2, *m*, and 2/*m*, there are 41 independent nonzero elements, consisting of

    3 elements with suffixes all equal,
    18 elements with suffixes equal in pairs,
    12 elements with suffixes having two *y*'s, one *x*, and one *z*,
    4 elements with suffixes having three *x*'s and one *z*,
    4 elements with suffixes having three *z*'s and one *x*.

Orthorhombic
For all three classes 222, *mm*2, and *mmm*, there are 21 independent nonzero elements, consisting of

    3 elements with suffixes all equal,
    18 elements with suffixes equal in pairs.

Tetragonal
For the three classes 4, $\bar{4}$, and 4*m*, there are 41 nonzero elements of which only 21 are independent. They are

$$xxxx = yyyy \qquad zzzz$$

$zzxx = zzyy$	$xyzz = \overline{yxzz}$	$xxyy = yyxx$	$xxxy = \overline{yyyx}$
$xxzz = yyzz$	$zzxy = \overline{zzyx}$	$xyxy = yxyx$	$xxyx = \overline{yyxy}$
$zxzx = zyzy$	$xzyz = \overline{yzxz}$	$xyyx = yxxy$	$xyxx = \overline{yxyy}$
$xzxz = yzyz$	$zxzy = \overline{zyzx}$		$yxxx = \overline{xyyy}$
$zxxz = zyyz$	$zxyz = \overline{zyxz}$		
$xzzx = yzzy$	$xzzy = \overline{yzzx}$		

For the four classes 422, 4*mm*, 4/*mmm*, and $\bar{4}$2*m*, there are 21 nonzero elements of which only 11 are independent. They are

$$xxxx = yyyy \qquad zzzz$$

$yyzz = zzyy$	$zzxx = xxzz$	$xxyy = yyxx$
$yzyz = zyzy$	$zxzx = xzxz$	$xyxy = yxyx$
$yzzy = zyyz$	$zxxz = xzzx$	$xyyx = yxxy$

Cubic
For the two classes 23 and *m*3, there are 21 nonzero elements of which only 7 are independent. They are

$$xxxx = yyyy = zzzz$$
$$yyzz = zzxx = xxyy$$
$$zzyy = xxzz = yyxx$$
$$yzyz = zxzx = xyxy$$
$$zyzy = xzxz = yxyx$$
$$yzzy = zxxz = xyyx$$
$$zyyz = xzzx = yxxy$$

TABLE 4.1 (*continued*)

For the three classes 432, $\bar{4}3m$, and $m3m$, there are 21 nonzero elements of which only 4 are independent. They are

$$xxxx = yyyy = zzzz$$
$$yyzz = zzyy = zzxx = xxzz = xxyy = yyxx$$
$$yzyz = zyzy = zxzx = xzxz = xyxy = yxyx$$
$$yzzy = zyyz = zxxz = xzzx = xyyx = yxxy$$

Trigonal

For the two classes 3 and $\bar{3}$, there are 73 nonzero elements of which only 27 are independent. They are

$$zzzz$$
$$xxxx = yyyy = xxyy + xyyx + xyxy \begin{cases} xxyy = yyxx \\ xyyx = yxxy \\ xyxy = yxyx \end{cases}$$

$$yyzz = xxzz \qquad xyzz = \overline{yxzz}$$
$$zzyy = zzxx \qquad zzxy = \overline{zzyx}$$
$$zyyz = zxxz \qquad zxyz = \overline{zyxz}$$
$$yzzy = xzzx \qquad xzzy = \overline{yzzx}$$
$$yzyz = xzxz \qquad xzyz = \overline{yzxz}$$
$$zyzy = zxzx \qquad zxzy = \overline{zyzx}$$

$$xxyy = \overline{yyyx} = yyxy + yxyy + xyyy \begin{cases} yyxy = \overline{xxyx} \\ yxyy = \overline{xyxx} \\ xyyy = \overline{yxxx} \end{cases}$$

$$yyyz = \overline{yxxz} = \overline{xyxz} = \overline{xxyz}$$
$$yyzy = \overline{yxzx} = \overline{xyzx} = \overline{xxzy}$$
$$yzyy = \overline{yzxx} = \overline{zxyx} = \overline{xzxy}$$
$$zyyy = \overline{zyxx} = \overline{zxyx} = \overline{zxxy}$$
$$xxxz = \overline{xyyz} = \overline{yxyz} = \overline{yyxz}$$
$$xxzx = \overline{xyzy} = \overline{yxzy} = \overline{yyxz}$$
$$xzxx = \overline{xyzy} = \overline{yxzy} = \overline{yzyx}$$
$$zxxx = \overline{zxyy} = \overline{zyxy} = \overline{zyyx}$$

For the three classes $3m$, $\bar{3}m$, and 32 there are 37 nonzero elements of which only 14 are independent. They are

$$zzzz$$
$$xxxx = yyyy = xxyy + xyyx + xyxy \begin{cases} xxyy = yyxx \\ xyyx = yxxy \\ xyxy = yxyx \end{cases}$$

$$yyzz = xxzz \qquad yyyz = \overline{yxxz} = \overline{xyxz} = \overline{xxyz}$$
$$zzyy = zzxx \qquad yyzy = \overline{yxzx} = \overline{xyzx} = \overline{xxzy}$$
$$zyyz = zxxz \qquad yzyy = \overline{yzxx} = \overline{xzyx} = \overline{xzxy}$$
$$yzzy = xzzx \qquad zyyy = zyxx = \overline{zxyx} = \overline{zxxy}$$
$$yzyz = xzxz$$
$$zyzy = zxzx$$

TABLE 4.1 (*continued*)

Hexagonal

For the three classes 6. $\bar{6}$, and 6/m, there are 41 nonzero elements of which only 19 are independent. They are

$$zzzz$$
$$xxxx = yyyy = xxyy + xyyx + xyxy \begin{cases} xxyy = yyxx \\ xyyx = yxxy \\ xyxy = yxyx \end{cases}$$

$$\begin{array}{ll} yyzz = xxzz & xyzz = \overline{yxzz} \\ zzyy = zzxx & zzxy = \overline{zzyx} \\ zyyz = zxxz & zxyz = \overline{zyxz} \\ yzzy = xzzx & xzzy = \overline{yzzx} \\ yzyz = xzxz & xzyz = \overline{yzxz} \\ zyzy = zxzx & zxzy = zyzx \end{array}$$

$$xxxy = \overline{yyyx} = yyxy + yxyx + xyyy \begin{cases} yyxy = \overline{xxyx} \\ yxyy = \overline{xyxx} \\ xyyy = \overline{yxxx} \end{cases}$$

For the four classes 622, 6mm, 6/mmm, and $\bar{6}m2$, there are 21 nonzero elements of which only 10 are independent. They are

$$zzzz$$
$$xxxx = yyyy = xxyy + xyyx + xyxy \begin{cases} xxyy = yyxx \\ xyyx = yxxy \\ xyxy = yxyx \end{cases}$$

$$\begin{array}{l} yyzz = xxzz \\ zzyy = zzxx \\ zyyz = zxxz \\ yzzy = xzzx \\ yzyz = xzxz \\ zyzy = zxzx \end{array}$$

Isotropic media

There are 21 nonzero elements of which only 3 are independent. They are

$$xxxx = yyyy = zzzz$$
$$yyzz = zzyy = zzxx \quad xxzz = xxyy = yyxx$$
$$yzyz = zyzy = zxzx = xzxz = xyxy = yxyx$$
$$yzzy = zyyz = zxxz = xzzx = xyyx = yxxy$$
$$xxxx = xxyy + xyxy + xyyx$$

---

[a] Each element is denoted by its subscripts in a Cartesian coordinate system with axes oriented along the directions of the principal crystalographic axes. A bar denotes the negative. (From Ref. [15], with permission.)

4.2 Symmetry Considerations

resonant) nonlinear susceptibility may be permuted separately from the polarization subscripts. This conditions is termed Kleinman symmetry; and is inherent in the Born–Oppenheimer approximation [16]. For an isotropic medium, Kleinman symmetry implies

$$\chi^{(3)}_{1122} = \chi^{(3)}_{1212} = \chi^{(3)}_{1221} = \tfrac{1}{3}\chi^{(3)}_{1111}. \tag{4.2.2}$$

Symmetry considerations also restrict the possible forms of the Raman tensors $\alpha^R_{\alpha\beta}$. In an isotropic material,

$$(\alpha^R_{11})^2 = (\alpha^R_{22})^2 = (\alpha^R_{33})^2, \tag{4.2.3}$$

$$\alpha^R_{11}\alpha^R_{22} = \alpha^R_{22}\alpha^R_{33} = \alpha^R_{33}\alpha^R_{11} = (1 - 2\rho_Q)(\alpha^R_{11})^2, \tag{4.2.4}$$

and

$$(\alpha^R_{12})^2 = (\alpha^R_{23})^2 = (\alpha^R_{13})^2 = \rho_Q(\alpha^R_{11})^2, \tag{4.2.5}$$

where ρ_Q is the usual Raman depolarization ratio. For "trace" modes $\rho_Q \ll \tfrac{3}{4}$ while for "totally depolarized" modes $\rho_Q = \tfrac{3}{4}$. The possible forms for α^R_{ij} for crystals and Raman modes of various symmetry [2] are given in Table 4.2.

Different combinations of chi-three and Raman tensor elements are accessed in different geometries. This fact and the variable number of distinguishable fields which result in a variable combinatorial factor D tend to confuse the comparison of geometries and techniques. Some common polarization conditions for CARS, RIKES/SRS, and four-wave mixing are shown in Tables 4.3 to 4.5 along with the relevant combinations of tensor elements.

Hellwarth has developed an elegant spherical tensor representation of chi-three for isotropic materials in the Born–Oppenheimer approximation [4]. The electronic background and (off-resonant) two-photon terms are grouped together as a real electronic hyperpolarizability σ

$$\sigma = 24(\chi^B_{1122} + \chi^T_{1122}) = 24(\chi^B_{1212} + \chi^T_{1212}) = 24(\chi^B_{1221} + \chi^T_{1221}). \tag{4.2.6}$$

The vibrational and orientational nuclear motions are parameterized in terms of two real, causal nuclear response functions, $a(t)$ and $b(t)$. The Fourier transform of these functions, $A(\Delta\omega)$ and $B(\Delta\omega)$, contribute directly to chi-three:

$$\begin{aligned}
24\chi^{(3)}_{\alpha\beta\gamma\delta}&(-\omega_s, \omega_0, \omega_1, -\omega_2) \\
&= \delta_{\alpha\beta}\delta_{\gamma\delta}[\sigma + 2A(\omega_1 - \omega_2) + B(\omega_0 + \omega_1) + B(\omega_0 - \omega_2)] \\
&+ \delta_{\alpha\gamma}\delta_{\beta\delta}[\sigma + 2A(\omega_0 - \omega_2) + B(\omega_0 + \omega_2) + B(\omega_1 - \omega_2)] \\
&+ \delta_{\alpha\delta}\delta_{\beta\gamma}[\sigma + 2A(\omega_0 + \omega_1) + B(\omega_1 - \omega_2) + B(\omega_0 - \omega_2)].
\end{aligned} \tag{4.2.7}$$

TABLE 4.2 Form of Raman Susceptibility Tensor $\alpha_{\alpha\beta}^R$ for Vibrational Modes of Various Symmetries in the 32 Crystal Classes[a]

System	Class		Raman Tensors		
Monoclinic	2 m 2/m	C_2 C_{1h} C_{2h}	$A(y)$ $A'(x,z)$ A_g $\begin{pmatrix} a & & d \\ & b & \\ d & & c \end{pmatrix}$	$B(x,z)$ $A''(y)$ B_g $\begin{pmatrix} & e & \\ e & & f \\ & f & \end{pmatrix}$	
Orthorhombic	222 mm2 mmm	D_2 C_{2v} D_{2h}	A $A_1(z)$ A_g $\begin{pmatrix} a & & \\ & b & \\ & & c \end{pmatrix}$	$B_1(z)$ A_2 B_{1g} $\begin{pmatrix} & d & \\ d & & \\ & & \end{pmatrix}$	$B_2(y)$ $B_1(x)$ B_{2g} $\begin{pmatrix} & & e \\ & & \\ e & & \end{pmatrix}$ $B_3(x)$ $B_2(y)$ B_{3g} $\begin{pmatrix} & & \\ & & f \\ & f & \end{pmatrix}$
Trigonal	3 $\bar{3}$	C_3 C_{3i}	$A(z)$ A_g $\begin{pmatrix} a & & \\ & a & \\ & & b \end{pmatrix}$	$E(x)$ E_g $\begin{pmatrix} c & d & e \\ d & -c & f \\ e & f & \end{pmatrix}$	$E(y)$ E_g $\begin{pmatrix} d & -c & -f \\ -c & -d & e \\ -f & e & \end{pmatrix}$
	32 3m 3m	D_3 C_{3v} D_{3d}	A_1 $A_1(z)$ A_{1g} $\begin{pmatrix} a & & \\ & a & \\ & & b \end{pmatrix}$	$E(x)$ $E(y)$ E_g $\begin{pmatrix} c & & \\ & -c & d \\ & d & \end{pmatrix}$	$E(y)$ $E(-x)$ E_g $\begin{pmatrix} & -c & -d \\ -c & & \\ -d & & \end{pmatrix}$
Tetragonal	4 $\bar{4}$ 4/m	C_4 S_4 C_{4h}	$A(z)$ A A_g $\begin{pmatrix} a & & \\ & a & \\ & & b \end{pmatrix}$	B $B(z)$ B_g $\begin{pmatrix} c & d & \\ d & -c & \\ & & \end{pmatrix}$	$E(x)$ $E(x)$ E_g $\begin{pmatrix} & & e \\ & & f \\ e & f & \end{pmatrix}$ $E(y)$ $E(-y)$ E_g $\begin{pmatrix} & & -f \\ & & e \\ -f & e & \end{pmatrix}$

Hexagonal

International	Schoenflies					
4mm	C_{4v}	$\begin{pmatrix} a & & \\ & a & \\ & & b \end{pmatrix}$	$\begin{pmatrix} c & & \\ & -c & \\ & & \end{pmatrix}$	$\begin{pmatrix} & d & \\ d & & \\ & & \end{pmatrix}$	$\begin{pmatrix} & & e \\ & & \\ e & & \end{pmatrix}$	$\begin{pmatrix} & & \\ & & e \\ & e & \end{pmatrix}$
422	D_4	$A_1(z)$	B_1	B_2	$E(x)$	$E(y)$
$\bar{4}2m$	D_{2d}	A_1	B_1	B_2	$E(-y)$	$E(x)$
4/mmm	D_{4h}	A_1	B_1	$B_2(z)$	$E(y)$	$E(x)$
		A_{1g}	B_{1g}	B_{2g}	E_g	E_g

6	C_6	$\begin{pmatrix} a & & \\ & a & \\ & & b \end{pmatrix}$	$\begin{pmatrix} c & d & \\ d & -c & \\ & & \end{pmatrix}$	$\begin{pmatrix} e & f & \\ f & -e & \\ & & \end{pmatrix}$	$\begin{pmatrix} & & \\ & & \\ & & \end{pmatrix}$ with entries $f, -e, -f$
$\bar{6}$	C_{3h}	$A(z)$	$E_1(x)$	$E_1(y)$	E_2	E_2
6/m	C_{6h}	A'	E''	E''	$E'(x)$	$E'(y)$
		A_g	E_{1g}	E_{1g}	E_{2g}	E_{2g}

622	D_6	$\begin{pmatrix} a & & \\ & a & \\ & & b \end{pmatrix}$	$\begin{pmatrix} & & c \\ & & \\ c & & \end{pmatrix}$	$\begin{pmatrix} & & \\ & & -c \\ & -c & \end{pmatrix}$	$\begin{pmatrix} d & & \\ & -d & \\ & & \end{pmatrix}$	$\begin{pmatrix} & d & \\ d & & \\ & & \end{pmatrix}$
6mm	C_{6v}	A_1	$E_1(x)$	$E_1(y)$	E_2	E_2
$\bar{6}m2$	D_{3h}	$A_1(z)$	$E_1(y)$	$E_1(-x)$	E_2	E_2
6/mmm	D_{6h}	A_1'	E''	E''	$E'(x)$	$E'(y)$
		A_{1g}	E_{1g}	E_{1g}	E_{2g}	E_{2g}

Cubic

23	T	$\begin{pmatrix} a & & \\ & a & \\ & & a \end{pmatrix}$	$\begin{pmatrix} b & & \\ & -b & \\ & & \end{pmatrix}$	$\begin{pmatrix} -b/\sqrt{3} & & \\ & -b/\sqrt{3} & \\ & & -2b \end{pmatrix}$ 		
m3	T_h	A	E	E		
432	O	A_g	E_g	E_g		
$\bar{4}3m$	T_d	A_1	E	E		
m3m	O_h	A_1	E	E		
		A_{1g}	E_g	E_g		

23	T	$\begin{pmatrix} & d & \\ d & & \\ & & \end{pmatrix}$	$\begin{pmatrix} & & d \\ & & \\ d & & \end{pmatrix}$	$\begin{pmatrix} & & \\ & & d \\ & d & \end{pmatrix}$		
m3	T_h	$F(x)$	$F(y)$	$F(z)$		
432	O	F_g	F_g	F_g		
$\bar{4}3m$	T_d	F_2	F_2	F_2		
m3m	O_h	$F_2(x)$	$F_2(y)$	$F_2(z)$		
		F_{2g}	F_{2g}	F_{2g}		

Note: the E-mode matrices in the cubic row are
$$\begin{pmatrix} b & & \\ & -b & \\ & & 0 \end{pmatrix} \quad \text{and} \quad \begin{pmatrix} -b/\sqrt{3} & & \\ & -b/\sqrt{3} & \\ & & -2b/\sqrt{3} \end{pmatrix}$$

[a] The coordinate system is referenced to the principal crystalographic axes. The direction of the dipole moment is given in parentheses for the polariton modes of acentric crystals. (From Ref. [2] with permission.)

TABLE 4.3

CARS/CSRS[a]		$P^{NL}(2\omega_1-\omega_2) = D\chi^{(3)}_{eff}E_1^2E_2^*$	
E_1	E_2	$D\chi^{(3)}_{eff}(-2\omega_1+\omega_2,\omega_1,\omega_1,-\omega_2)$	$(\alpha^R_{eff})^2$
1 ←	←	$3\chi^{(3)}_{1111}$	$2(\alpha^R_{11})^2$
2 ←	↑	$3\chi^{(3)}_{1221}$	$2(\alpha^R_{12})^2$
3 ←[b]	←[b]	$\frac{3}{2}[\chi^{(3)}_{1111}+2\chi^{(3)}_{1122}+\chi^{(3)}_{1221}]$	$[(\alpha^R_{11})^2+2\alpha^R_{11}\alpha^R_{22}+(\alpha^R_{12})^2]$
4 ↑	↑	$3\chi^{(3)}_{1122}$	$[(\alpha^R_{12})^2+\alpha^R_{11}\alpha^R_{22}]$
5		$3[\chi^{(3)}_{1111}\cos^2\theta\cos\phi+\chi^{(3)}_{1221}\sin^2\theta\cos\phi]$ $-\chi^{(3)}_{1122}\sin 2\theta\sin\phi]$	$2(\alpha^R_{11})^2\cos^2\theta\cos\phi+2(\alpha^R_{12})^2\sin^2\theta\cos\phi$ $-[(\alpha^R_{12})^2+\alpha^R_{11}\alpha^R_{22}]\sin 2\theta\sin\phi$

[a] P's always vertical.
[b] Along [1,1,0] crystal axis.

TABLE 4.4

SRS/RIKES[a]		$P^{NL}(-\omega_2) = D\chi^{(3)}_{eff}|E_1|^2 E_2$	
E_1	E_2	$D\chi^{(3)}_{eff}(\omega_2, -\omega_1, \omega_1, \omega_2)$	$(\alpha^R_{eff})^2$
6 ↑	↑	$6\chi^{(3)}_{1111}$	$(\alpha^R_{11})^2$
7 →	↑	$6\chi^{(3)}_{1221}$	$(\alpha^R_{12})^2$
8 ↗	→	$3[\chi^{(3)}_{1122} + \chi^{(3)}_{1212}]$	$\frac{1}{2}[\alpha^R_{11}\alpha^R_{22} + (\alpha^R_{12})^2]$
9 ↻	→	$3i[\chi^{(3)}_{1122} - \chi^{(3)}_{1212}]$	$\frac{1}{2}[\alpha^R_{11}\alpha^R_{22} - (\alpha^R_{12})^2]$

[a] P's always vertical.

TABLE 4.5

4 WM[a]			$P^{NL}(\omega_0 + \omega_1 - \omega_2) = D\chi^{(3)}_{eff} E_0 E_1 E_2^*$	
E_0	E_1	E_2	$D\chi^{(3)}_{eff}(-\omega_0 - \omega_1 + \omega_2, \omega_0, \omega_1 - \omega_2)$	$(\alpha^R_{eff})^2$
10 ↑	↑	↑	$6\chi^{(3)}_{1111}$	$(\alpha^R_{11})^2$
11 →	→	↑	$6\chi^{(3)}_{1221}$	$(\alpha^R_{12})^2$
12 ↑	→	→	$6\chi^{(3)}_{1122}$	$\alpha^R_{22}\alpha^R_{11}{}^b$, $(\alpha^R_{12})^{2\,c}$
13 →	↑	→	$6\chi^{(3)}_{1212}$	$(\alpha^R_{12})^{2\,b}$, $\alpha^R_{11}\alpha^R_{22}{}^c$
14 ↻	↻	→	$3i[\chi^{(3)}_{1122} - \chi^{(3)}_{1212}]$	$\pm\frac{1}{2}[\alpha^R_{11}\alpha^R_{22} - (\alpha^R_{12})^2]^d$
15 →	↻	↻	$3i[\chi^{(3)}_{1212} - \chi^{(3)}_{1221}]$	0^b, $\frac{1}{2}[\alpha^R_{11}\alpha^R_{22} - (\alpha^R_{12})^2]^c$
16 ↗ϕ	↗θ	→	$3[\chi^{(3)}_{1212}\sin\theta\cos\phi - \chi^{(3)}_{1122}\cos\theta\sin\phi]$	$[(\alpha^R_{12})^2\sin\theta\cos\phi - \alpha^R_{11}\alpha^R_{22}\cos\theta\sin\phi]^e$

[a] P's always vertical.
[b] Resonant at $\omega_1 - \omega_2 = \Omega_Q$.
[c] Resonant at $\omega_0 - \omega_2 = \Omega_Q$.
[d] Plus sign for $\omega_1 - \omega_2 = \Omega_Q$, minus sign for $\omega_0 - \omega_2 = \Omega_Q$.
[e] For $\omega_1 - \omega_2 = \Omega_Q$, for $\omega_0 - \omega_2 = \Omega_Q$ interchange θ and ϕ.

In the following section, the imaginary parts of $A(\Delta\omega)$ and $B(\Delta\omega)$ are related to the cross sections for polarized and depolarized scattering. The parameters $A(0)$ and $B(0)$ fulfill

$$A(0) = \int_0^\infty \frac{\operatorname{Im} A(\omega)}{\omega} d\omega, \qquad B(0) = \int_0^\infty \frac{\operatorname{Im} B(\omega)}{\omega} d\omega \qquad (4.2.8)$$

and if all modes are depolarized, $A(0) = -B(0)/3$.

4.3 RELATIONSHIP BETWEEN χ^R AND THE SPONTANEOUS CROSS SECTION

The magnitude of the Raman contributions to chi-three can be related to the cross section for spontaneous Raman scattering. For the Lorentzian modes previously considered, the Raman tensors can be related to the total Stokes cross sections

$$\rho_{gg}|\alpha^R_{\alpha\beta}|^2 = (\lambda/2\pi)^4 \, d\sigma_{\alpha\beta}/d\Omega, \qquad (4.3.1)$$

where polarized scattering occurs for $\alpha = \beta$ and depolarized scattering for $\alpha \neq \beta$; ρ_{gg} is the diagonal density matrix element for the lower Raman level [2]. For isotropic materials, the differential Raman cross sections can be related directly to the imaginary part of chi-three [4, 12],

$$\operatorname{Im}\chi^{(3)}_{1111}(\omega_2, -\omega_1, \omega_1, -\omega_2) = \frac{\pi c^4}{24\hbar\omega_1\omega_2^3} \frac{d^2\sigma_{11}}{d\Omega\, d(\omega_1 - \omega_2)} (e^{-\hbar(\omega_1 - \omega_2)/kT} - 1)$$

$$= \operatorname{Im}[A(\omega_1 - \omega_2) + B(\omega_1 - \omega_2)]/12 \qquad (4.3.2)$$

$$\operatorname{Im}\chi^{(3)}_{1221}(\omega_2, -\omega_1, \omega_1, -\omega_2) = \operatorname{Im}\chi^{(3)}_{1212}(\omega_2, -\omega_1, \omega_1, -\omega_2) = \frac{\pi c^4}{24\hbar\omega_1\omega_2^3}$$

$$\times \frac{d^2\sigma_{12}}{d\Omega\, d(\omega_1 - \omega_2)} (e^{-\hbar(\omega_1 - \omega_2)/kT} - 1)$$

$$= \operatorname{Im} B(\omega_1 - \omega_2)/24. \qquad (4.3.3)$$

4.4 WAVE-VECTOR MATCHING

In Section 2.9 it was shown how to relate the nonlinear polarization to the detected signal in a variety of spectroscopic techniques. The generation of a detectable signal requires achieving a wave-vector matching condition between the driven polarization and a corresponding freely propagating wave. While this condition is automatically fulfilled in stimulated Raman

4.4 Wave-Vector Matching

gain spectroscopy, in other cases, achieving this phase matching condition may restrict the geometrical configuration.

When the input frequencies and wave vectors are $\omega_0, \omega_1, -\omega_2$ and $\mathbf{k}_0, \mathbf{k}_1, \mathbf{k}_2$, respectively, the output frequency and the wave vector of the polarization will be $\omega_s = \omega_0 + \omega_1 - \omega_2, \mathbf{k}_p = \mathbf{k}_0 + \mathbf{k}_1 - \mathbf{k}_2$. Phase matching requires that the quadrilateral formed by $\mathbf{k}_0, \mathbf{k}_1, \mathbf{k}_2$, and $\mathbf{k}_s$ close, at least approximately (see Fig. 4.4). Algebraically,

$$\Delta k = |\mathbf{k}_0 + \mathbf{k}_1 - \mathbf{k}_2 - \mathbf{k}_s| \to 0. \tag{4.4.1}$$

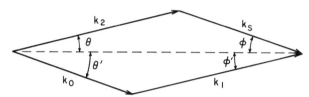

Fig. 4.4 The general wave-vector matching condition in coherent Raman spectroscopy. In CARS $\theta' = \phi' = 0$. "Folded" boxcars can be represented by creasing the page along the dotted line.

In coherent anti-Stokes Raman spectroscopy, there are three distinct frequencies and wave vectors interacting. Dispersion of the index of refraction of the medium makes perfect phase matching impossible for collinear waves, even in an isotropic medium. Expanding the index of refraction as a power series in the frequency shift $\Delta = \omega_1 - \omega_2$,

$$n(\omega_1 - \Delta) = n_1 + n'_1 \Delta + \tfrac{1}{2} n''_1 \Delta^2, \tag{4.4.2}$$

we find that (4.4.1) can be fulfilled best when the incident beams cross at an angle θ given by

$$\theta = \left[\left(\frac{n''_1}{n_1} + \frac{2n'_1}{n_1 \omega_1}\right)\left(1 + \frac{\omega_1 - \omega_2}{\omega_1}\right)\right]^{1/2} (\omega_1 - \omega_2). \tag{4.4.3}$$

Clearly achieving optimum phase matching at each frequency in a Raman spectrum requires varying the angle with the laser frequency [2]. In gases, the dispersion may be weak enough that the coherence length for collinear beams

$$l_c = \pi/\Delta k = [\pi c/(\omega_1 - \omega_2)^2]/(2n'_1 + 2\omega_1 n''_1) \tag{4.4.4}$$

is longer than the sample for every frequency shift. In that case, the decreased interaction length that results from a nonzero crossing angle makes perfect phase matching undesirable.

Another strategy for coping with dispersion is tight focusing of collinear beams [17]. In this case, each beam contains a range of wave vectors, some combination of which approximately fulfills the wave-vector matching condition at each frequency difference.

In birefringent media, the phase-matching condition can be quite subtle, and it is recommended that exact calculations be performed using the best available index data except when all beams are polarized parallel to one another and parallel to a single crystallographic axis. In accentric crystals, the Raman frequency itself will vary with direction and the wave-vector difference $\mathbf{k}_1 - \mathbf{k}_2$. The spectra observed in this case contain information on the spatial dispersion of the polariton frequencies [2, 18].

The presence of an additional input beam relaxes some of the geometrical constraints in four-wave mixing. Again, the correct crossing angles are best determined by trigometric calculations using the quadrilateral in Fig. 4.4 and the best available index of refraction data. In "boxcars," two input waves have the same frequency but different direction. In "submarine," the frequency differences $\omega_1 - \omega_2$ and $\omega_s - \omega_0$ are small, and the wave-vector matching condition becomes nearly as unrestrictive as in the stimulated Raman gain case, when the beam pairs with nearby frequencies propagate collinearly (see Fig. 1.11).

4.5 COHERENT ANTI-STOKES RAMAN SPECTROSCOPY

Coherent anti-Stokes Raman spectroscopy (CARS) is the most widely practiced nonlinear mixing spectroscopy technique [3]. In CARS, two beams at frequency ω_1 and ω_2 are mixed in the sample to generate a new frequency: $\omega_s = 2\omega_1 - \omega_2$. If ω_1 corresponds to the laser frequency in a spontaneous scattering experiment and ω_2 to the Stokes-scattered photon, the output occurs at the corresponding anti-Stokes frequency. If $\omega_1 < \omega_2$, the analogous technique is called coherent Stokes Raman spectroscopy (CSRS).

The major experimental advantage of CARS—and most coherent Raman techniques—is the large signal produced. In a typical liquid or solid, the effective third-order nonlinear susceptibility is approximately 10^{-13} esu, and in a typical CARS experiment, the laser powers are 10 kW or so and the interaction length is 0.1 cm. Under these circumstances, (2.9.9) and (4.1.10) imply a CARS output power of $\sim 1\text{W}$, while conventional Raman scattering would give collected signal power of $\sim 100\ \mu\text{W}$ with the same lasers. Since the CARS output is directional, the collection angle can be five orders of magnitude smaller than that needed in spontaneous scattering. Taken together, these two factors imply that CARS is nine orders of magnitude less

4.5 Coherent Anti-Stakes Raman Spectroscopy

sensitive to sample fluorescence than spontaneous scattering. The advantage is actually somewhat greater; since the output is at a higher frequency than the input, Stokes' law of fluorescence implies that spectral filtering can further reduce the background fluorescence level.

The main disadvantages of CARS are (1) an unavoidable electronic background nonlinearity that alters the lineshapes and can limit the detection sensitivity; (2) a signal that scales as the square of the spontaneous scattering signal (and as the cube of laser power), making the signals from weakly scattering samples difficult to detect; and (3) the need to fulfill the phase matching requirements of Section 4.4. While other techniques avoid these difficulties, CARS remains the most popular coherent Raman technique.

4.5.1 CARS Experiments in Liquids and Solids

Many different kinds of lasers can be used as the sources of the two input beams. The main requirements are good beam quality, adequate power, and tunability. Figure 4.5 shows a system that uses two Hänsch-type dye lasers simultaneously pumped by a single nitrogen laser [19]. The peak power of these dye lasers can be in the range of 10 to 100 kW at about 10 Hz repetition rate with a linewidth of approximately 0.5 cm^{-1}. The beam divergence is typically twice that of a TEM$_{00}$ mode. The second harmonic of a pulsed Q-switched Nd:YAG laser can be used as one of the CARS input beams and also as the pump for the tunable laser. The power produced by such a system is much greater than that of a nitrogen laser-based system ($P_1 \gtrsim 5$ MW $P_2 \gtrsim 0.1$ MW), but too much power is not entirely useful. Generally, 100-kW peak power or less is sufficient for CARS in liquids and solids; greater power can lead to self-focusing, stimulated Raman oscillation, and damage to the sample. To date, only Ahkmanov has successfully employed cw lasers in a CARS experiment in liquids [20].

A lens with a focal length between 10 and 20 cm focuses the two parallel incident beams into the sample. Care must be taken to ensure that the two beam waists lie at the same distance from the lens. Chabay et al. have devised a knife-edge test procedure to ensure that the beam waists coincide spatially [19].

The positions of mirrors M_2 and M_3 in Fig. 4.5 must be adjusted so that the CARS wave-vector matching condition is fulfilled. If these mirrors are mounted on translation stages oriented so that translation of the mirrors does not affect the orientation of the plane defined by the crossing beams, the crossing angle can be adjusted without greatly affecting the beam overlap. In Carreira's CARS microanalysis system, the mirror positions and angles are automatically optimized by computer [21].

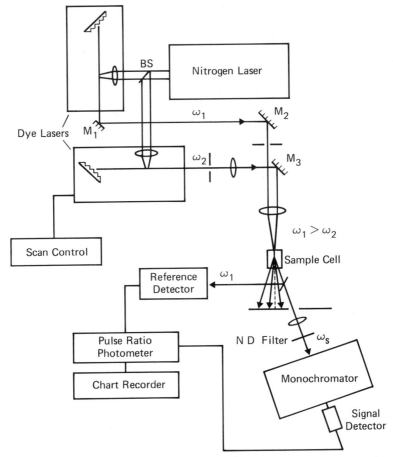

Fig. 4.5 Experimental apparatus for CARS. The two parallel dye laser beams are focused into a sample at the crossing angle θ. The CARS output is selected spatially and then spectrally, finally being detected and averaged by a photodiode and a ratioing gated integrator (from Ref. [19]).

The anti-Stokes output beam is generated at approximately the crossing angle of the input beams. A moveable knife edge or iris can block the laser beams while allowing the CARS signal into the detection system. A lens which images the interaction region in the sample onto the detector will collect the CARS signal beam radiated at any angle. Alternatively, the entire detection system can be pivoted around the sample, with the collection angle optimized by computer.

A monochromator in the detection system to spectrally filter the output and eliminate scattering of the incident frequencies is shown in Fig. 4.5.

4.5 Coherent Anti-Stokes Raman Spectroscopy

A colored glass filter or an interference filter will suffice for the strongest and highest frequency modes.

The output beam can be detected by a PIN photodiode or an inexpensive photomultiplier. Because the CARS signal can span a dynamic range of eight decades, it is convenient to provide a calibrated variable attenuator to control the intensity at the detector surface. The electrical pulses produced by the detector must be integrated and amplified, and then averaged using a gated integrator or computer system. A second reference detector which monitors and averages the intensity of the pump laser is shown in Fig. 4.5. Some of the noise resulting from fluctuations in the laser output can be eliminated by dividing the CARS signal by this monitor signal. Other workers have monitored the CARS outputs from two different samples in order to reduce the noise due to laser fluctuations [22].

The CARS spectrum is actually the normalized and averaged signal intensity plotted as a function of difference frequency $\omega_1 - \omega_2$. In practice, ω_1 is often fixed while ω_2 is varied by tuning the dye laser grating. It should be noted that the signal frequency ω_s varies with ω_2 and care should be taken that the signal frequency ω_s remain within the bandpass of the spectral filtering system.

In condensed phases, the nonresonant contribution to the third-order nonlinear susceptibility can be relatively large and the CARS line shape reflects the destructive interference between the Raman contribution to the nonlinear susceptibility and the nonresonant contribution. As a function of the frequency difference $\omega_1 - \omega_2$, the CARS intensity for an isolated Lorentzian Raman modes varies as

$$I(\omega_1 - \omega_2) \propto |\chi^R + \chi^{NR}|^2$$

$$\propto \frac{\left\{\Omega_Q^2 - (\omega_1 - \omega_2)^2 + \dfrac{\mathcal{N}\Omega_Q \alpha_{\text{eff}}^{R2}}{2Dh\chi_{\text{eff}}^{NR}}\right\}^2 + (\omega_1 - \omega_2)^2/T_2^2}{[\Omega_Q^2 - (\omega_1 - \omega_2)^2]^2 + 4(\omega_1 - \omega_2)^2/T_2^2}. \quad (4.5.1)$$

The frequency difference between the maximum and minimum reflects the ratio of the Raman tensor to the nonresonant susceptibility. The line-shape function in Eq. (4.5.1) plotted in Fig. 4.3a, while Fig. 4.6 shows a corresponding experimental result, the CARS signal in the vicinity of the optical phonon mode in diamond obtained with various polarization conditions [22]. The varying positions of the minimum reflects different ratios of resonant to nonresonant susceptibility found for different combinations of the nonlinear susceptibility tensor elements.

Another complication occurs in absorbing samples because of the attenuation of the incident and signal amplitudes which depend, in general,

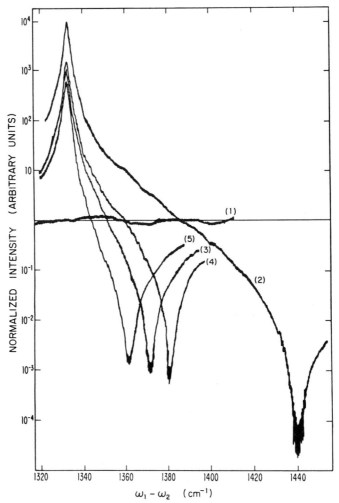

Fig. 4.6 CARS Spectra of diamond. Each of the traces corresponds to one of the polarized conditions of Table 4.3 (from Ref. [22]).

upon the frequencies and intensities. If the attenuation coefficients are independent of intensity, the amplitudes of the incident waves decrease as

$$E_j \propto e^{-\kappa z/2}.$$

The sample length yielding the maximum CARS signal amplitude is

$$l_{opt} = 2 \left| \frac{\ln[\kappa_s/(2\kappa_1 + \kappa_2)]}{2\kappa_1 + \kappa_2 - \kappa_s} \right|. \tag{4.5.2}$$

4.5 Coherent Anti-Stokes Raman Spectroscopy

If all of the attenuation coefficients are equal, (4.5.2) implies an optimum sample density of 0.48. Near this attenuation level, small changes in the absorption coefficient for one of the beams does not significantly affect the CARS spectrum [2].

4.5.2 CARS in Gases: Pulsed Laser Techniques

The nonlinear susceptibility of a gas at STP is 100 times smaller than that of a typical liquid, but the interaction length can be much longer since (4.4.3) implies a near-zero crossing angle for optimal phase matching. Since the optical breakdown threshold for gases is orders of magnitude higher than for liquids, all of the power produced by a Nd:YAG based CARS laser system can be profitably employed. Thus, the CARS signals can be nearly as large as in liquids [3].

The typical experimental setup for pulsed CARS in gases is very similar to that in Fig. 4.5, the main differences being a longer sample cell and some provision for making the input beams collinear. High intensity beams can be combined using a dichroic beam splitter, a carefully aligned dispersing prism, or—if the planes of polarization are orthogonal—a glan-laser prism. The signal beam is best separated from the incident beam using a prism or interference filter rather than an easily damaged monochromator.

4.5.3 Multiplex CARS

Multiplex CARS is an attractive alternative to the single-frequency pulsed techniques for gas spectroscopy [3, 20]. In multiplex CARS, the Stokes beam at ω_2 is produced by a broad band laser. Since the CARS output is linear in the Stokes input, the CARS signal due to each frequency component of the broad-band laser can be separated by a spectrograph and detected with an optical multichannel analyzer or a photographic plate. The resolution is obviously limited by the spectrograph employed, but much of the Raman spectrum can be obtained in a single laser shot. Averaging over many shots gives spectra of the same quality as single-channel CARS spectra if the lasers have comparable output powers.

A multiplex CARS spectrum of N_2 taken in an atmospheric pressure methane/air flame using an Nd:YAG laser as the pump source is shown in Fig. 4.7. These researchers report that CARS spectra in laboratory flames (0.1–1.0 atm total pressure) can be recorded with minimum spectral slit widths [23]. They also demonstrated that the single pulse experiments are feasible even in the sub-Torr pressure range by taking a CARS spectrum of CH_4 at 600 mTorr pressure.

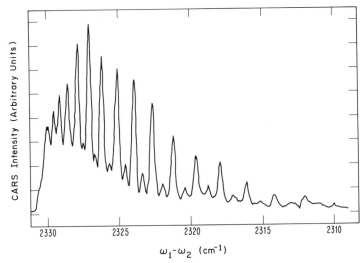

Fig. 4.7 Multiplex CARS spectrum of N_2 in a laboratory flame. Fitting the strengths of the individual lines of this resolved Q branch reveals the rotational temperature (from Ref. [23]).

It is also possible to take multiplex CARS spectra from condensed phases. In particular, when the sample length is very short as in one-photon resonant CARS experiments, the phase-matching requirement can be relaxed and a broad band laser can be employed for the Stokes beam.

4.5.4 cw CARS

High resolution (3 MHz or 10^{-4} cm^{-1}) can be achieved in gas-phase CARS spectra with single-mode cw lasers. The obvious disadvantage is the drastically reduced CARS signal level due to the low power levels of cw lasers. Thus, the observations of cw CARS signal has been limited to relatively strong Raman transitions.

The first cw CARS experiment was reported by Barrett and Begley [24]. Their CARS signal level at the peak of the Q branch of the v_1 mode of methane was actually lower than that obtained by spontaneous scattering under the same conditions. More recently, Byer et al. have observed Raman lines in H_2, D_2, and CH_4 using the cw CARS apparatus depicted in Fig. 4.8 [25]. An argon ion laser (5145 Å) and a cw dye laser (6054 Å) operating in single modes at power levels of 650 and 10 mW, respectively, were collinearly overlapped in the high pressure (27 atm) gas cell. The two beams were focused with a 10.5-cm lens to minimum radii of 6 and 20 μm for the argon and dye laser, respectively. The CARS signal due to the $Q(v_1)$ branch of CH_4 at 2416.7 cm^{-1} (or a Q-branch line of H_2 or D_2) was isolated from

4.6 Raman-Induced Kerr Effect Spectroscopy

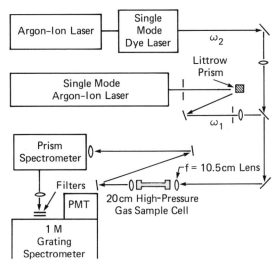

Fig. 4.8 CW CARS apparatus for ultrahigh resolution Raman spectroscopy. In a gas sample, optimum phase matching requires collinear propagation. Separation of the weak CARS output from background signals due to laser fluorescence is accomplished spectroscopically. In more recent experiments, a supersonic molecular beam replaced the high-pressure gas sample (from Ref. [25]).

laser radiation and transmitted through a prism prefilter, spike filters, and a 1-m grating spectrometer and then finally detected with a cooled photomultiplier tube operating in the photon counting mode.

By employing a similar apparatus to that shown in Fig. 4.8, Fabelinsky et al. have obtained the high-resolution CARS spectrum of C_2H_2 and D_2 at various pressures between 0.06 and 40 atm. Instead of a spectrometer, they used a four-prism filter system which allowed a reduction of the pump laser light intensity by more than 17 orders of magnitude while transmitting more than 50% of the anti-Stokes signal beam [26].

4.6 RAMAN-INDUCED KERR EFFECT SPECTROSCOPY

A strong pump wave propagating through a nonlinear medium induces an intensity-dependent dichroism and birefringence which can alter the polarization of a weaker probe beam. If the probe beam is initially linearly polarized, the change in polarization can be detected as an increase in the intensity transmitted through a crossed polarizer. The nonlinear susceptibility that produces this optical Kerr effect has contributions from the Raman modes of the material in addition to contributions from reorientation of molecules and electrons in the medium. The Raman terms make their major contribution to the Kerr effect when the pump and probe frequencies differ by a Raman frequency, and the detection of these resonances is the basis of Raman induced Kerr effect spectroscopy (RIKES) [7, 9, 12].

The polarization conditions used in RIKES are numbered 8 and 9 in Table 4.4. The Raman resonant terms in the nonlinear susceptibilities in Table 4.4 are clearly comparable to the CARS resonances. The main advantages of RIKES are (1) a phase-matching condition that is automatically fulfilled for every propagation direction and frequency combination (in isotropic media), and (2) the suppression of the nonlinear background when the pump wave is circularly polarized as in polarization condition 9. The main disadvantage is a background signal due to stress-induced birefringence in the sample and optics.

A typical RIKES setup is shown in Fig. 4.9. The lasers used are essentially identical to those recommended for CARS in the previous sections, but in RIKES the probe is the laser that need not be tunable [2]. A glan-laser prism ensures that the probe beam is linearly polarized and the beam is focused into a sample cell by a lens of roughly 15-cm focal length. After the sample cell, the probe beam is refocused into a Babinet Soleil variable wave plate which partially compensates for depolarization due to the sample or optics. The probe beam is then blocked by a second carefully aligned

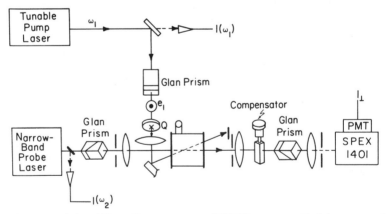

Fig. 4.9 Raman-induced Kerr effect spectroscopy (RIKES) apparatus. Both lasers are pulsed and are often pumped by the same N_2 laser source. The glan prisms ensure that the beams are linearly polarized. The quarter-wave plate at Q can change the pump wave from linear to circular polarization. The Babinet Soleil compensator and the glan prism polarization analyzer are adjusted to prevent the probe polarization from entering the Spex monochromator. The photomultiplier labeled PMT responds only to pump laser induced changes in the probe polarization. The Raman-induced Kerr effect is in many ways analogous to the polarization spectroscopy discussed in Chapter 3 (from Ref. [2]).

4.6 Raman-Induced Kerr Effect Spectroscopy

glan prism and the RIKES signal beam is focused into a monochromator and a detector. Judiciously placed irises provide spatial filtering.

The polarization of the tunable laser is controlled by means of a linear polarizer and a rotatable quarter wave plate Q. The pump beam is focused into the cell and overlapped with the probe as in CARS, but because there is no wave-vector matching condition to fulfill, the angle between the beams can be made very small and the length of the interaction region maximized. When the pump is linearly polarized, the nonresonant background susceptibility is sufficient to provide a RIKES signal whenever the beams overlap. If the quarter-wave plate is adjusted to give circularly polarized light in the sample, that nonresonant background will disappear. RIKES spectra of benzene taken by scanning the pump frequency in such a system with a linear and a circular polarized pump [2] are shown in Fig. 4.10.

As in polarization spectroscopy, stress-induced birefringence in the sample and optics produces a troublesome background signal. The Babinet Soleil compensator can eliminate the effects of uniform stress, but not the effects

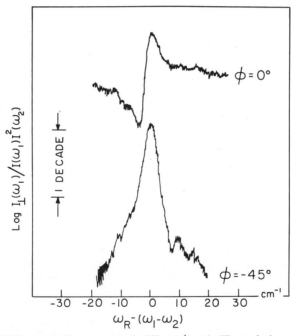

Fig. 4.10 RIKES spectra of benzene near the 992 cm^{-1} mode. The angle ϕ corresponds to the setting of the wave plate Q in Fig. 4.9. At $\phi = 0$, the pump is linearly polarized and a strong background signal results from the nonresonant Kerr effect. At $\phi = -45°$, the nonresonant background is suppressed leaving a Lorentzian peak with structure in the wings due to isotope shifts (from Ref. [2]).

of variations in the stress. To minimize the birefringence background signal, the compensator plate and polarization analyzer must be sequentially re-adjusted to reach a minimum. With care, the birefringence background intensity can be reduced to less than 10^{-6} of the probe intensity, or two orders of magnitude less than a typical RIKES signal.

Multiplex RIKES is possible when the pump laser is narrow band and the probe broad band. A spectrograph separates the frequency component of the beam that is transmitted through the polarization analyzer and the RIKES spectrum can then be recorded on an OMA or photographic plate. Since the phase matching condition is automatically fulfilled, multiplex RIKES spectra can be taken over a wide frequency range in liquids, solids, and gases [12].

4.6.1 Optical Heterodyne Detected RIKES

The sensitivity of Raman-induced Kerr effect spectroscopy is radically improved and the importance of birefringence background signals is reduced when optical heterodyne techniques are employed to detect the signal amplitude [2, 7]. The resulting technique—termed OHD–RIKES—has greater demonstrated sensitivity than any other coherent Raman technique, and yet is remarkably simple in operation. The signals obtained scale linearly with the concentration and Raman cross section, as does the spontaneous scattering intensity, and the line shapes are directly comparable to those in spontaneous scattering.

One form of OHD–RIKES apparatus is diagrammed in Fig. 4.11. The pump laser is pulsed or modulated, and the probe laser is generally cw and stable. In Fig. 4.11, the probe beam is shown to be slightly elliptically polarized. The horizontal component E_O acts as a local oscillator field, which is out of phase by $\pi/2$ with the vertical component which corresponds to the linear "probe" beam of ordinary RIKES. In this configuration, the polarization analyzer transmits the horizontal component of the probe beam as well as the RIKES amplitude E_R. In Fig. 4.11, the RIKES amplitude is shown as a pulse having the same shape as the pump laser pulse and lying on top of the local oscillator. Spatial and spectral filters separate these signal amplitudes from scattered pump light, and the corresponding intensity

$$I(t) = \frac{nc}{8\pi} |\mathbf{E}_O(t) + \mathbf{E}_R(t)|^2 = I_O(t) + I_R(t) + \frac{n_s c}{4\pi} \operatorname{Re}[\mathbf{E}_O^*(t) \cdot \mathbf{E}_R(t)] \quad (4.6.1)$$

is detected photoelectrically. An electrical bandpass filter separates the ac signals due to $E_R(t)$ from the near dc signals due to $I_O(t)$. Because $E_O \gg E_R$, the heterodyne term $(n_s c/4\pi) \operatorname{Re}[\mathbf{E}_O^*(t) \cdot \mathbf{E}_R(t)]$ makes the dominant ac con-

4.6 Raman-Induced Kerr Effect Spectroscopy

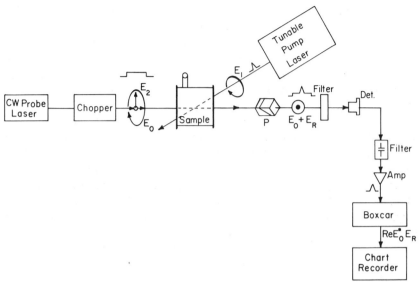

Fig. 4.11 Schematic of optically heterodyne detected Raman-induced Kerr effect spectroscopy (OHD–RIKES) apparatus. The probe laser has a stable output power, but is chopped to prevent heating of the sample. The probe polarization has a strong component labeled E_2 which acts as the RIKES probe and a much weaker orthogonal component E_0 that acts as a local oscillator in the heterodyne detection scheme. The pump laser alters the probe polarization via the optical Kerr effect, the amplitude transmitted through the polarization analyzer as a result of this effect is labeled E_R. Electronic filtering separates the short pulse due to the Raman-induced Kerr effect from the longer pulse due to the local oscillator (from Ref. [2]).

tribution to the electrical signal. The heterodyne term is then averaged and plotted as a function of the pump laser frequency.

This heterodyne signal is proportional to the spontaneous scattering signal, but it is also sensitive to the phase of $\chi^{(3)}$. When the probe wave is elliptically polarized as in Fig. 4.11, the local oscillator is in quadrature with the probe and the detected lineshape corresponds to $\mathrm{Re}\, D\chi^{(3)}_{\mathrm{eff}}$ as given in Table 4.4. When the pump is circularly polarized, this choice of phases gives a resonance-type line shape similar to spontaneous scattering, except that polarized and depolarized peaks will have opposite signs. If the pump is linearly polarized, dispersion-type line shapes will appear. A local oscillator in phase with the probe can be obtained when the output of the probe laser is linearly polarized and the polarization analyzer P is rotated away from the angle of minimum probe transmission. In this case, the spectrum reflects $\mathrm{Im}\, D\chi^{(3)}_{\mathrm{eff}}$ as in Table 4.4 and a linearly polarized pump gives the resonance line shapes while a circularly polarized pump gives dispersion type line shapes.

The optimum strength of the local oscillator wave depends upon the characteristics of the lasers and detectors. The details of this optimization will be discussed in Section 4.10. For the poorly stabilized 0.5-W single mode cw argon laser used as a probe in our laboratory, the optimum local oscillator power was 0.3 mW. Laser noise degraded the OHD–RIKES signal when the local oscillator level was larger, and shot noise degraded the spectra when the level was lower [7].

The high-resolution capabilities of OHD–RIKES with cw lasers is illustrated in Fig. 4.12. This cw OHD–RIKES spectrum of the $Q(1)$ $v = 0 \to 1$ line of hydrogen was taken by Owyoung using apparatus similar to Fig. 4.11 except that a single-mode argon laser was used as a pump and a cw dye laser became the probe [6]. Note that both resonance and dispersion line shapes can be obtained. The slight asymmetry in the Im $\chi^{(3)}$ line shape can be eliminated by modulating the pump polarization rather than the pump

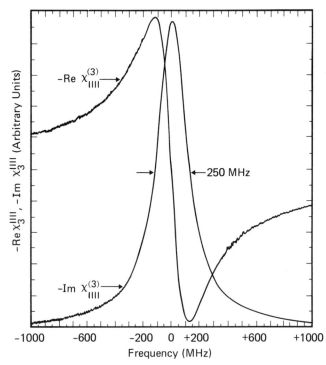

Fig. 4.12 A high-resolution gas phase OHD–RIKES spectrum due to Owyoung. The lasers used were cw with the pump chopped to produce a modulation detectable with a lock-in amplifier. Both phases of the Raman resonance are clearly visible (from Ref. [6]). Experimental parameters: 2.5-atm hydrogen. $Q_{01}(1)$ transition 23°C, 450-mW pump power, 15-mW probe power, and 300-ms time constant.

4.7 Stimulated Raman Gain and Loss Spectroscopy

intensity. Since the OHD–RIKES signal scales as the product of two laser intensities, rather than three, it is better suited than CARS for low-power cw experiments. At low densities, the cw OHD–RIKES signal level is orders of magnitude larger than that of CARS, and the sensitivity is correspondingly better.

4.7 STIMULATED RAMAN GAIN AND LOSS SPECTROSCOPY

The modern techniques of stimulated Raman spectroscopy (SRS) employ stable cw probe lasers and detect the small ($\sim 10^{-5}$) changes in intensity due to Raman gain or loss induced by a pump laser [6]. SRS shares with OHD–RIKES the advantages of a signal linearly proportional to the spontaneous cross section (and to the product of two laser intensities), and an automatically fulfilled phase-matching condition. It has the additional advantage of being insensitive to depolarization. The main disadvantage is that SRS is much more sensitive to laser noise than OHD–RIKES [2].

The high-resolution SRS system developed by Owyoung et al. [27] is shown in Fig. 4.13. The pump source was a single mode Ar$^+$ laser operating

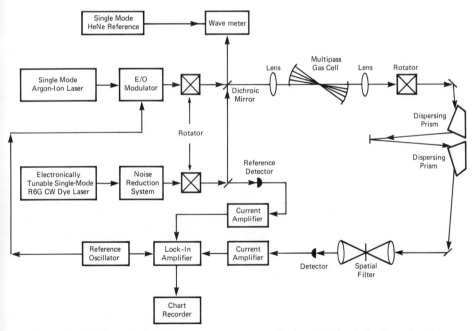

Fig. 4.13 A high-resolution Raman spectrometer employing stimulated Raman gain with cw lasers (from Ref. [27]).

at 5145 Å. It's output was modulated electrooptically and was combined with the probe on a dichroic mirror. The probe laser was a feedback stabilized cw dye laser that was capable of tuning over 1 cm^{-1} with a resolution of 1 MHz. An electrooptic demodulator suppressed the characteristic power fluctuations, and a reference detector allowed the remaining fluctuations to be subtracted from the SRS signal by a differential input lock-in amplifier.

The coincident beams were directed into a multipass cell. After 97 passes through the sample, the SRS signal had been enhanced by a factor of 50 over the single-pass case, but the probe intensity had been reduced to 0.5 mW. The output beams were separated by dispersing prisms and a spatial filter, and the modulation of the probe due to the stimulated Raman effect was detected by a PIN diode and differential lock-in amplifier with quantum-noise-limited sensitivity.

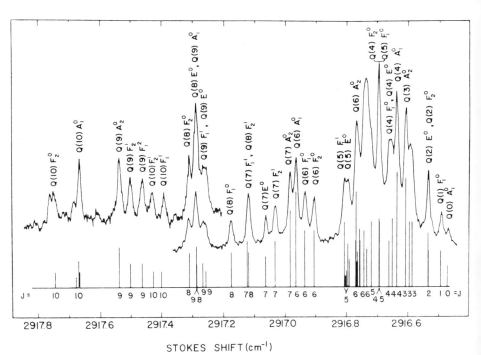

Fig. 4.14 Stimulated Raman gain spectrum of the Q branch of the v_1 fundamental of methane taken with the instrument in Fig. 4.13. The linewidths shown correspond to the Doppler widths expected for the Raman transition. The labels identify the transitions by angular momentum and symmetry (from Ref. [27]). Experimental parameters: 35-Torr methane, 650-mW pump power, 1.5-mW probe power, 3-s time constant, and 23.5°C.

The most striking demonstration of the resolution and sensitivity of this cw SRGS system is shown in Fig. 4.14. The spectrum displays the v_1 fundamental mode of methane (near 2917 cm^{-1}) under the pressure of 35 Torr. This spectrum gives the first fully resolved Q-branch Raman spectrum of a polyatomic molecule.

The sensitivity of a hybrid SRS system employing a pulsed pump laser as in Fig. 4.11 can be expected to be 1000 times greater than the all-cw system in Fig. 4.13. Such systems may prove useful for coherent Raman spectroscopy in gases, liquids, and solids.

4.8 FOUR-WAVE MIXING

When three laser beams at three different frequencies, ω_0, ω_1, and ω_2 are mixed in the sample by the third-order nonlinear susceptibility, the coherently generated nonlinear signal can have the frequency $|\omega_s| = |\omega_0 + \omega_1 - \omega_2|$ [2]. Raman resonances occur when $|\omega_0 - \omega_2| = \Omega_Q$ and when $|\omega_1 - \omega_2| = \Omega_Q$. The process producing the output at $|\omega_s| = |\omega_0 + \omega_1 - \omega_2|$ reduces to the CARS processes when $\omega_0 \to \omega_1$, and to the RIKES or SRS process when $\omega_0 \to -\omega_1$.

Four-wave mixing (4 WM) techniques permit the suppression of troublesome background signal which limit the sensitivity of the corresponding three-wave mixing technique: the nonresonant electronic background signal in case of CARS and the strain-induced birefringence in case of RIKES. Since the nonlinear susceptibility tensors that apply to 4 WM have simultaneous resonances at more than one Raman frequency, it is possible to compare two different Raman cross sections by direct interference of the susceptibilities. In some applications, these advantages outweigh the difficulties inherent in an experiment employing three synchronized, overlapped, and phase matched laser beams and the difficulties connected with the low signal levels that result when the background signal is suppressed. CARS-type processes also produce outputs at $2\omega_1 - \omega_2$ and $2\omega_0 - \omega_2$ which must be suppressed by poor phase matching or separated from the 4 WM output spatially or spectrally.

The kind of two-dimensional spectrum that can be obtained by scanning $\omega_0 - \omega_2$ and $\omega_1 - \omega_2$ over a region where two components of a mixture have Raman modes [28] is shown in Fig. 4.15. Fitting the line shapes obtained in this way normalizes the cross sections of the two modes to one another. The effective nonlinear susceptibility and Raman tensor elements that are accessible in some of the possible 4 WM polarization conditions are shown in Table 4.5. The most interesting are conditions (14)–(16) where some degree of nonlinear background suppression is feasible [2].

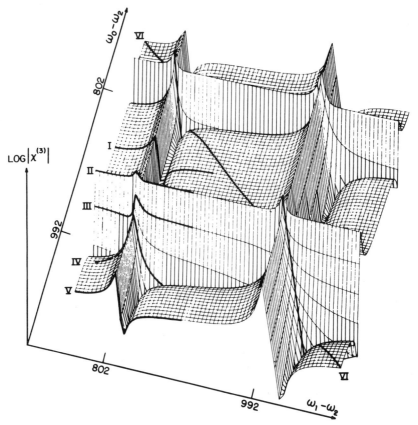

Fig. 4.15 Two-dimensional plot of the dispersion of $|\chi^{(3)}_{1111}|^2$ versus $\omega_1 - \omega_2$ and $\omega_0 - \omega_2$ in a 1:1 mixture of benzene and cyclohexane. The heavy lines show the regions explored experimentally by four-wave mixing; a CARS trace would correspond to path VI. The ratio of that Raman cross sections of benzene and cyclohexane was determined from such plots to be 3.45:1 (from Ref. [28]).

4.9 APPLICATIONS

It is not feasible even to list all the applications and experimental results obtained to date using coherent Raman spectroscopy. Such a task has been performed by a number of admirable articles that have recently appeared [2–8, 23]. What follows is an abbreviated discussion of proven applications with some typical examples cited for each.

4.9 Applications

4.9.1 Combustion Diagnostics: Concentration and Temperature Measurement

The coherent Raman signals can be easily separated from luminescent backgrounds. This property of the coherent Raman techniques have made them especially useful in combustion diagnostics where luminescence often overwhelms the spontaneous scattering signal. In an early experiment, J. P. Taran was able to map the distribution of H_2 formed by pyrolysis in a methane–air flame with a spatial resolution of a few millimeters [8].

The spatial resolution of CARS can be enhanced in two ways: if the ω_1 and ω_2 beams cross at a small angle, the length of the active region where the beams overlap is greatly reduced. So too is the anti-Stokes signal which exits from the interaction region at an angle between the two incident waves, but the signals are often so large that this reduction is unimportant. The "boxcars" technique of Fig. 1.11 also limits the length of the overlap region at the expense of signal intensity, but the angles involved can be larger than in the former scheme [29].

The Raman term in the nonlinear susceptibility depends upon the difference in populations of the two coupled levels. At thermal equilibrium

$$w^e = \rho_{rr} - \rho_{gg} = Z^{-1}(e^{-E_r/kT} - e^{-E_g/kT}), \qquad (4.9.1)$$

where Z is the partition function for the sample species. By comparing the coherent Raman signal strengths for several transitions with known values of α^R, the temperature can be in principle determined. True thermal equilibrium, however, often does not exist in plasmas and flames, and a distinction must be made among rotational, vibrational, and translational temperatures. When the fundamental and hot bands of a Raman transition are well resolved, the vibrational temperature can be readily assigned. The rotational temperature can also be unequivocally inferred if individual lines can be resolved [2].

Temperature and concentration data have been obtained by Roh et al. in multiplex CARS experiments, some performed upon the exhaust of a jet engine. The multiplex techniques can obtain all the information necessary to assign a temperature in a single laser shot, and are thus more applicable to the study of turbulent or rapidly evolving systems [23].

Some of the difficulties in extracting concentration and temperature information from CARS spectra can be avoided in OHD–RIKES and SRS. In these techniques, the coherent spectrum can be made to reproduce the spontaneous scattering spectrum and previously proven techniques of analysis applied.

4.9.2 Raman Cross-Section and Nonlinear Susceptibility Measurements

Measurements of the total cross section for spontaneous scattering are among the most difficult experiments in Raman spectroscopy. Accurate corrections must be made for the collection geometry, detector quantum efficiency, spectrometer transfer function, etc. It is thus not surprising that few such measurements exist with expected uncertainty less than 10%.

In contrast, the precise measurement of the stimulated Raman gain or loss is relatively straightforward in a cw or quasi-cw experiment. For collinearly focused identical TEM_{00} pump and probe beams, the change in probe power due to Raman gain upon traversing a single focus is

$$\mathscr{P}_H(\omega_2)/\mathscr{P}_1\mathscr{P}_2 = 96\pi^2 \omega_1 \omega_2/c^2 \, \text{Im} \, \chi^{(3)}_{\alpha\beta\beta\alpha}(\omega_2, -\omega_1, \omega_1, -\omega_2), \quad (4.9.2)$$

which can be related to the total cross section [6]. To determine the gain and cross section directly, three powers must be measured accurately, one of them rather small. While satisfactory techniques exist for doing so, it is often simpler to compare the Raman gain for one sample with that of a well-known material such as benzene. Using an interferometric variation of SRS, Owyoung and Peercy measured the peak gain of the benzene 992 cm^{-1} mode in terms of the well-known optical Kerr constant of CS_2. Their result,

$$\mathscr{N}(\alpha^R)^2 T_2/12\hbar = 31.8 \pm 2.2 \times 10^{-14} \quad \text{esu}, \quad (4.9.3)$$

agrees with the most accurate determinations by spontaneous scattering [6, 22].

Coherent Raman techniques can also provide accurate spectroscopic measurements of the non-Raman contributions to the third-order nonlinear susceptibility χ^{NR}, and the need for such measurements provided much of the motivation for much of the early development of CARS. In CARS, 4 WM, and RIKES with a linearly polarized pump, the Raman line shape of an isolated mode has a maximum and a minimum as in Fig. 4.3. The frequency difference between maximum and minimum depends upon the ratio of the background susceptibility to the Raman matrix element squared

$$\Delta\omega = \{4T_2^{-2} + [\mathscr{N}(\alpha^R_{\text{eff}})^2/4\hbar D\chi^{NR}_{\text{eff}}]^2\}^{1/2}, \quad (4.9.4)$$

where α^R_{eff} and $D\chi^{NR}_{\text{eff}}$ depend on the specific polarization and frequency condition. This ratio can thus be obtained without the need for potentially uncertain power or intensity measurements.

Some of the combinations of $\chi^{(3)}$ tensor elements that are accessible in different polarization and frequency conditions are given in Tables 4.3 to 4.5. Each combination of tensor elements gives a different frequency difference between maximum and minimum, as is shown in Fig. 4.6. In a

4.9 Applications

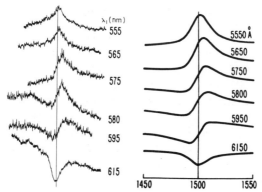

Fig. 4.16 Resonant CARS spectra of cyanoalbumin (vitamin B_{12}) in water solution for various pump wavelengths. The Raman frequency is 1500 cm^{-1}. Resonance effects alter the line shape, making maxima into minima, but the effects observed at moderate intensity can be described by Eq. (2.8.9) as shown by the calculated curves (from Ref. [2]).

series of such experiments, the Raman and non-Raman contributions to each tensor element can thus be determined.

The non-Raman terms in $\chi^{(3)}$ have scientific and technological importance. For example, the term responsible for self-focusing is

$$n_2 = (12\pi/n)\chi^{(3)}_{1111}(-\omega,\omega,-\omega,\omega). \qquad (4.9.5)$$

In liquids, glasses, and cubic crystals,

$$\chi^{(3)}_{1111}(-\omega,\omega,-\omega,\omega) = \chi^{NR} = \tfrac{1}{8}\sigma + \tfrac{1}{6}[A(0) + B(0)],$$

and thus a measurement of the background susceptibilities in CARS, RIKES, and interferometric SRS can completely characterize n_2.

Once the ratio of the Raman and nonlinear background contributions to chi-three have been determined for one mode, the ratio of the cross sections for other modes can be obtained by repeating the experiment at different values of $\omega_1 - \omega_2$. These techniques work well for materials with strong, well-characterized Raman lines and for mixtures when proper account is taken of local field effects. The nonlinear susceptibilities of materials without such modes can be determined using a "sandwich sample" [30].

Two-photon absorption contributes additional resonances in $\chi^{(3)}$. When the two-photon absorption lines are narrow and Lorentzian, the resonances have the same general line shape as the Raman resonances, but with different frequency arguments. By comparing the dispersion of $\chi^{(3)}$ due to the two-photon resonances with that due to known Raman resonances, various authors have normalized two-photon absorption cross sections to Raman cross sections [22, 31].

4.9.3 High-Resolution Molecular Spectroscopy

Perhaps the most dramatic advantage of coherent Raman spectroscopy over conventional Raman scattering is in resolution and frequency precision. In coherent Raman spectroscopy with cw sources, an instrumental resolution of 0.0001 cm^{-1} or 3 MHz can readily be obtained, and interferometric wavelength determination techniques with a frequency precision of ± 10 MHz are also available. Since the resolution is built into the laser sources, the signal levels obtained are as large as for less well-engineered cw sources. Recently, well-stabilized cw lasers have been used as oscillators and amplified to tens of kilowatts in laser-pumped dye laser amplifiers. The resolution and frequency precision of coherent Raman spectrometers employing such lasers can be estimated as 20 and ± 50 MHz, respectively.

At these levels of resolution and precision, the Doppler effect gives the largest contribution to the linewidth and to the experimental uncertainty in the transition frequency. For typical vibrational frequencies, the Doppler width is of order 100 MHz. Even so, coherent Raman spectroscopy offers a potential improvement of two orders of magnitude in the precision of rotational and vibrational constants.

A remarkable early result is shown in Fig. 4.14: the resolved Q branch of methane at 35 Torr. Similar data is now available for a number of diatomic and polyatomic molecules [27].

4.9.4 Raman Spectra of Fluorescent and Resonant Samples

By working with pulsed lasers, gated detection, judiciously prepared samples, and carefully chosen (or oscillating) wavelengths, Raman spectroscopists have obtained spectra of fluorescent samples. Spontaneous scattering techniques must fail, however, when the cross section for prompt fluorescence exceeds the cross section for spontaneous scattering by several orders of magnitude. In these cases, the coherent Raman techniques remain viable. Fluorescence can be suppressed by spatial and spectral filtering, and the coherent Raman signal enhanced by heterodyne detection. Particularly in CARS, fluorescence-free spectra can be easily obtained, since the signal frequency is at the anti-Stokes side of the input laser beams. In addition, CRS offers particular advantages for Raman studies of biologically important materials without the danger of sample degradation since ~ 1 mW of average power is sufficient to take CRS spectra under resonant conditions [21].

Systematic application of resonantly enhanced CARS was first reported by Hudson *et al.* who studied diphenyl octatetraene dissolved in benzene. They observed enhancement of the CARS intensity without significant distortion of the line shapes. Drastic line-shape changes were observed by

4.9 Applications

Nestor *et al.* from their CARS spectra of vitamin B_{12} in water. A series of CARS traces taken a different laser wavelengths along with theoretical plots obtained by Lynch *et al.* using Eq. (2.8.9) [2] is shown in Fig. 4.16. The Raman resonant structure no longer necessarily appears as a peak but can evolve into an "inverted peak" or destructive interference between the resonant susceptibility of the solute species and the "background" susceptibility. Detailed studies of CARS resonant excitation profiles were reported by Carriera *et al.* with several biological compounds. By employing a computerized CARS microanalysis system, CARS detection of sample concentrations as low as $\sim 5 \times 10^{-7}$ M was possible with β-carotene in benzene [32].

When the absorption lines are narrow, additional resonances appear in the CRS spectra due to processes similar to those diagrammed in Fig. 2.8. In SRS and RIKES, some of the resonances result from changes in the steady-state populations of excited electronic levels, and some can have linewidths narrower than the Doppler width [6]. These points are illustrated in Fig. 4.17. A near-resonant SRS spectrum of I_2 vapor taken with a cw argon laser pump and a cw dye laser probe is shown in Fig. 4.17a. The cross sections of the two peaks are enhanced by the nearby P(13) and R(15) 43-0 X → B absorption lines. The lines are Gaussian with width 44 MHz (HWHM). When the pump is tuned onto resonance, the Raman peaks break up into narrower lines corresponding to individual hyperfine transitions with differing resonant velocity groups [2]. Somewhat analogous effects have also been seen in the resonant CARS spectrum of I_2.

For resonant work, CARS and 4 WM analogs have real advantages over SRS, OHD–RIKES. In the latter two techniques, thermal blooming and excited state absorption induced by the pump alters the local oscillator intensity at the detector, producing spurious signals and noise. Also, the depolarization ratio in the resonant case is often near $\rho_Q = \frac{1}{3}$ resulting in a suppression of the Raman signal whenever RIKES or 4-WM-type background suppression techniques are employed. All the coherent Raman techniques demonstrated to date are transmission techniques. For opaque and translucent samples, conventional spontaneous scattering continues to be the best available technique.

4.9.5 Polariton Dispersion: Spectroscopy in Momentum Space

In noncentrosymmetric media, the frequencies of the polariton modes depend dramatically upon the wave vector. In spontaneous scattering, the wave vector of the polariton depends upon the difference of the wave vectors of the incident and scattered radiation: $\mathbf{q} = \mathbf{k}_I - \mathbf{k}_S$. Unfortunately, spontaneous scattering techniques must collect light over a finite range of wave

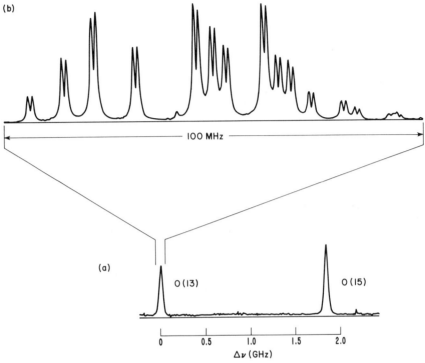

Fig. 4.17 Resonant stimulated Raman spectra of I_2. Trace (a) shows the O(13) and O(15) lines of the $v'' = 0 \rightarrow 9$ band as they appear with the 5145 Å pump laser tuned 1 GHz from the P(13) R(15) 43-0 rovibronic absorption line. The two peaks are Doppler broadened. Trace (b) shows the additional structure that results when the pump is tuned into resonance with the P(13) absorption line. Each pair of lines corresponds to amplification of the probe by a different inverted hyperfine transition. The splitting between the two lines of a pair corresponds to the one-photon Rabi frequency. The overall hyperfine structure can be compared to Fig. 3.21. Related effects appear under resonant conditions in other CRS techniques (from Ref. [2]).

vector $\Delta \mathbf{k}_S$ in order to have a finite signal level, and the uncertainty in the collected wave vector introduces an uncertainty in the wave vector of the polariton.

Since coherent Raman techniques employ diffraction limited beams propagating in definite directions, the wave vector of the excited polariton can be better defined than in spontaneous scattering, at least in principle. In fact, the usual practice of working with tightly focused beams in order to obtain the maximum signal from a finite length sample significantly reduces the precision with which $\mathbf{q}$ can be specified. Nevertheless, a number of workers have employed CARS and 4 WM techniques to refine polariton disperison curves [2, 18].

Coffinet and De Martini originated spectroscopy of this type using stimulated Raman oscillators. Since their frequencies were fixed, they plotted the output signal as a function of the crossing angles of their beams. Today, their experiment would be considered an example of "boxcars," but they termed the technique "spectroscopy in k space" because the wave vector of the nonlinearly excited polariton, but not the frequency, varied as the crossing angle was scanned. Ultimately, using various pairs of input frequencies, De Martini succeeded in plotting out the lower branch of the transverse polariton curve of GaP with high accuracy [33].

At the surfaces of noncentrosymmetric crystals, there exist surface polariton modes with frequencies between the LO and TO bulk polaritons. These modes have been studied by spontaneous scattering, but they can also be driven coherently by frequency mixing techniques. De Martini *et al.* have also plotted out the dispersion curve and linewidth for these modes in GaP using their k-space spectroscopy version of "boxcars" [34]. With due care, the driven surface polariton can also be coupled out the sample and detected directly. The resonances observed in this case are more properly assigned to $\chi^{(2)}$ than to $\chi^{(3)}$.

4.9.6 Vibrational and Rotational Relaxation Measurements

When the optical fields used to drive a Raman mode are turned off, Eq. (4.1.4) implies that the coordinate for that mode continues to oscillate until dephasing and relaxation processes damp it out. The time scale involved ranges from picoseconds for most liquids and solids to many nanoseconds for low-pressure gases. The dephasing and relaxation times can be measured using short pulsed lasers in the time domain, or by analyzing the CRS line shape in the frequency domain.

Kaiser and his co-workers pioneered the time domain techniques using picosecond lasers and stimulated Raman oscillation in the sample to drive the vibrational mode. A delayed picosecond probe laser sampled the coherent and incoherent part of the excitation at a later time. In this way, the vibrational relaxation and dephasing times of liquids such as N_2, CCl_4, and CH_3OH were measured [35].

Lee and Ricard have been studing vibrational relaxation using a transient version of boxcars. In this technique, picosecond pulses from a doubled Nd:glass laser and a dye laser are overlapped in space and time to drive the vibrational mode of the sample, and a delayed pulse from the glass laser is used as a probe. Since each of the waves propagates in a slightly different direction, the phase matching diagram in Fig. 4.4 predicts a definite direction for the CARS pulse produced by the delayed probe. Vibrational relaxation can be studied in this way at intensity levels well below the threshold for stimulated Raman oscillation [35].

Similar studies have been performed by Heritage using a picosecond version of stimulated Raman gain spectroscopy [36]. Because of the advantages of heterodyne detection, Heritage can obtain vibrational relaxation data using synchronously pumped cw-mode locked dye lasers even in reflection.

4.10 JUDGING THE MERITS: THE SIGNAL-TO-NOISE RATIO

The coherent Raman techniques produce much more intense optical signals than spontaneous scattering, but the increased signal level does not necessarily imply better sensitivity. Features intrinsic to the nonlinear mixing process and noise phenomena common to most laser sources impact the signal-to-noise ratio. The formalism of signal-to-noise analysis developed for communications and applied to coherent Raman spectroscopy points out the strengths and various weaknesses of various techniques [2, 7].

The starting point for such an analysis is the nonlinear signal amplitude defined in Eq. (2.9.7). Only part of this amplitude results from the Raman process, the rest coming from background terms in the nonlinear susceptibility. The true coherent Raman signal amplitude $\mathbf{E}_R$ is obtained from (2.9.7), when the proper coherent Raman polarization $\mathbf{P}^Q$ from (4.1.8) is substituted for $\mathbf{P}^{NL}$.

At the photodetector surface, there are other amplitudes besides the Raman signal amplitude. The amplitudes that add coherently to the Raman amplitude can be added together to form the local oscillator amplitude $\mathbf{E}_{LO}$. In CARS, this amplitude results from the nonlinear background susceptibility, while in SRS it is the transmitted probe laser amplitude. The intensity at the photodetector surface thus is

$$I(x, y) = \frac{nc}{8\pi} |\mathbf{E}_R(x, y) + \mathbf{E}_{LO}(x, y)|^2 + I_B(x, y)$$

$$= I_{LO}(x, y) + I_R(x, y) + I_B(x, y) + \frac{nc}{4\pi} \text{Re}[\mathbf{E}_{LO}^*(x, y) \cdot \mathbf{E}_R(x, y)], \quad (4.10.1)$$

where $I_{LO}(x, y)$, $I_R(x, y)$, and $I_B(x, y)$ are the local oscillator, coherent Raman, and incoherent background intensities, respectively [see (2.9.12) for comparison]. The total detector response is proportional to the integral of this intensity over the photodetector surface

$$\rho = \lambda \int I(x, y) \, dx \, dy. \quad (4.10.2)$$

4.10 Judging the Merits: The Signal-to-Noise Ratio

The factor λ defines the dimensions of that detector response. It is convenient to set λ equal to the quantum efficiency of the detector, and in this case the detector response has the dimensions of power. It is also convenient to separate the various contributions to the detector response

$$\rho = \rho_{LO} + \rho_R + \rho_B + \rho_H, \quad (4.10.3)$$

where each term on the right of (4.10.3) results from the corresponding term on the right of (4.10.1).

The total response due to the coherent Raman Signal is $\rho_H + \rho_R$, and for the purpose of this argument both are assumed positive. The heterodyne response ρ_H and the Raman intensity response ρ_R can be related to the corresponding signal powers: $\rho_H = 2(\rho_R \rho_{LO})^{1/2}$.

Fluctuations in the laser power and mode structure produce noise on each of the terms in the detector response. The mean square noise fluctuation due to laser fluctuations is

$$\langle \delta\rho_L^2 \rangle = \langle \delta\rho_{LO}^2 \rangle + \langle \delta\rho_B^2 \rangle + \langle \delta\rho_R^2 \rangle + \langle \delta\rho_H^2 \rangle \quad (4.10.4)$$

and

$$\langle \delta\rho_\sigma^2 \rangle = \varepsilon_\sigma \rho_\sigma^2, \quad (4.10.5)$$

where the fraction due to laser power fluctuation is

$$\varepsilon_\sigma = (1/\rho_\sigma^2) \sum_k (\partial\rho_\sigma/\partial\mathscr{P}_k)^2 \langle \delta\mathscr{P}_k^2 \rangle \quad (4.10.6)$$

and $\langle \delta\mathscr{P}_k^2 \rangle$ is the mean square fluctuation in the power of the kth laser, within the bandwith of the detection system.

The shot noise that results from the quantized nature of light detection can be expressed in terms of a mean square fluctuation

$$\langle \delta\rho_q^2 \rangle = 2\rho_1 \rho = 2\rho_1(\rho_{LO} + \rho_R + \rho_B + \rho_H), \quad (4.10.7)$$

where $\rho_1 = \hbar\omega_s \Delta\nu$ is the response produced by the detection of a single photon. The nonoptical noise sources—thermal noise, radio-frequency interference, dark-current fluctuations, etc.—can be expressed in terms of the noise equivalent input power ζ

$$\langle \delta\rho_{NO}^2 \rangle = \zeta^2 \Delta\nu. \quad (4.10.8)$$

All of these noise terms are defined for a given detection bandwidth $\Delta\nu$.

The signal-to-noise can then be expressed in general as

$$\frac{S}{N} = \frac{2(\rho_R \rho_{LO})^{1/2} + \rho_R}{\{\langle \delta\rho_L^2 \rangle + \langle \delta\rho_q^2 \rangle + \langle \delta\rho_{NO}^2 \rangle\}^{1/2}}. \quad (4.10.9)$$

Equation (4.10.9) is the key result, from it can be estimated the signal-to-noise ratio for any laser spectroscopy experiment.

This formalism can be modified somewhat and applied to spontaneous scattering (COORS). In the spontaneous case, there is no local oscillator, so $\rho_{LO} \equiv 0$. The remaining intensity response is proportional to the incident laser power, Raman cross section, concentration, sample length, and collection efficiency λ_c,

$$\rho_{rs} = \lambda \lambda_c \mathcal{N} l (d^2\sigma/d\,\Delta\omega\,d\Omega) \mathcal{P}_L \qquad (4.10.10)$$

while the corresponding intensity response in the coherent Raman case depends upon the *square* of the cross section and concentration. This difference in the scaling law is a fundamental disadvantage for coherent Raman spectroscopy for samples at low concentration or with weak scattering cross sections.

Various authors have employed this sort of formalism to delineate the capabilities of various real and proposed coherent Raman systems [2, 6, 7, 37]. A plot due to Eesley of the signal-to-noise ratio expected for his laser system plotted as a function of the concentration of a Raman active species [7] is shown in Fig. 4.18. Also plotted are the quantum-noise-limited signal-to-noise ratios of the coherent Raman techniques and of spontaneous scattering. Because of photomultiplier dark current and residual sample luminescence, the practical sensitivity of spontaneous scattering falls significantly below the theoretical limit at low concentrations.

Even so, the sensitivity of spontaneous scattering exceeds that of all the coherent Raman techniques except OHD–RIKES. Increased laser power and greater stability would improve the performance of all of the coherent Raman techniques, with CARS benefiting more from increased power, and SRS from increased stability. A fair representation of the state of the art in 1980 is, however, given by Fig. 4.18. At high concentrations, the signal-to-noise levels are limited entirely by laser stability, while at low concentrations the limit is set for these CRS experiments by a combination of shot noise and laser stability. Similar comments apply to polarization spectroscopy and other nonlinear laser technologies.

The increased complexity of coherent Raman technology does not lead to an increased sensitivity in comparison to spontaneous scattering. If an experiment can be done in a straightforward way using spontaneous scattering, it should be done that way. If not, it may be reasonable to choose one of the coherent Raman techniques. The technique providing the best sensitivity for samples without significant absorption is optical heterodyne detected Raman-induced Kerr effect spectroscopy. Equivalent results can be obtained using stimulated Raman gain or loss when the shot noise at the detector exceeds the intensity fluctuation noise. CARS and CSRS should be used in resonant cases where excited state absorption and other effects produce unacceptable background for SRS or OHD–RIKES. The other techniques are curiosities best avoided.

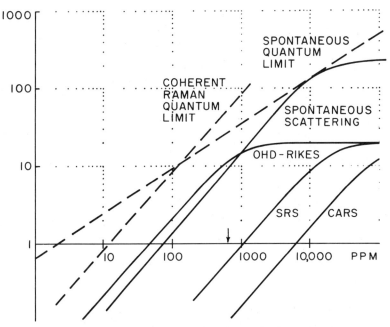

Fig. 4.18 Calculated signal-to-noise ratios for three forms of coherent Raman spectroscopy plotted as a function of sample concentration of benzene in CCl_4. Also plotted are the quantum-noise-limited signal-to-noise ratios for all CRS techniques and for spontaneous scattering. The curve for spontaneous scattering represents the performance presently achieved in practice. The CRS signal to noise ratios are calculated for a single laser shot for a system with 0.5-W probe power and a 50-kW pump laser with 0.5 and 5.0% rms fluctuations, respectively (from Ref. [2]).

REFERENCES

1. G. Placzek, "Marx Handbuch der Radiologie", 2nd ed., Vol. VI, pp. 20–374. Akademische Verlagsgesellschaft, Leipzig, 1934.
2. M. D. Levenson and J. J. Song, Coherent Raman spectroscopy, *in* "Coherent Nonlinear Optics" (M. S. Feld and V. S. Letokhov, eds.) (Topics in Current Physics 21), pp. 293–373. Springer-Verlag, Berlin, 1980.
3. J. W. Nibler and G. V. Knighten, Coherent anti-Stokes Raman spectroscopy, *in* "Raman Spectroscopy of Gases and Liquids" (A. Weber, ed.) (Topics in Current Physics 11). Springer-Verlag, Berlin, 1978.
4. R. W. Hellwarth, Third order nonlinear susceptibility of liquids and solids, *Progr. Quant. Electron.* **5**, 1 (1977).
5. S. A Akhmanov, Coherent active spectroscopy of combinatorial (Raman) scattering with tunable oscillators; comparison with the spontaneous scattering technique, *in* "Nonlinear Spectroscopy" (N. Bloembergen, ed.), pp. 217–254. North-Holland Publ., Amsterdam, 1977.
6. A. Owyoung, CW stimulated Raman spectroscopy, *in* Chemical Applications of Nonlinear Raman Spectroscopy" (A. B. Harvey, ed.). Academic Press, New York, 1981.

7. G. L. Eesley, "Coherent Raman Spectroscopy." Pergamon, Oxford, 1981.
8. S. Druet and J.-P. Taran, Coherent anti-Stokes Raman spectroscopy, in "Chemical and Biochemical Applications of Lasers" (B. Moore, ed.). Academic Press, New York, 1978.
9. P. D. Maker and R. W. Terhune, *Phys. Rev.* **137**, A801 (1965).
10. E. J. Woodbury and W. E. Ng, *Proc. IRE* **50**, 2367 (1962).
11. W. T. Jones and B. P. Stoicheff, *Phys. Rev. Lett.* **13**, 657 (1964).
12. D. Heiman, R. W. Hellwarth, M. D. Levenson, and G. Martin, *Phys. Rev. Lett.* **34**, 189 (1976).
13. N. Bloembergen, *Amer. J. Phys.* **35**, 989–1023 (1967).
14. Y. R. Shen, in "Light Scattering in Solids" (M. Cardona, ed.) (Topics in Appl. Physics 8), pp. 275–328. Springer-Verlag, Berlin, 1975.
15. P. N. Butcher, Nonlinear optical phenomena, Eng. Bull. 200, Ohio State Univ. Press, 1965.
16. D. A. Kleinman, *Phys. Rev.* **129**, 1977 (1962).
17. G. C. Bjorklund, *IEEE J. Quant. Electron.* **QE-11**, 287 (1975).
18. C. Flytzanis, in "Quantum Electronics" (H. Rabin and C. L. Tang, eds.), Vol. 1. Academic Press, New York, 1975.
19. I. Chabay, G. K. Klauminzer, and B. S. Hudson, *Appl. Phys. Lett.* **28**, 27 (1976).
20. S. A. Akhmanov, A. F. Bunkin, S. G. Ivanov, N. I. Koroteev, A. I. Kovrigin, and I. C. Shumay, in "Tunable Lasers and Applications" (A. Mooradian, T. Jaeger, and P. Stokseth, eds.), p. 389. Springer-Verlag, Berlin, 1976.
21. L. A. Carreira, L. P. Goss, and T. B. Malloy, *J. Chem. Phys.* **68**, 280 (1978).
22. M. D. Levenson and N. Bloembergen, *Phys. Rev. B* **10**, 4447 (1974).
23. A. B. Harvey, "Chemical Applications of Nonlinear Raman Spectroscopy". Academic Press, New York, 1981.
24. J. J. Barrett and R. F. Begley, *Appl. Phys. Lett.* **27**, 129 (1975).
25. M. A. Henesian, L. Kulevskii, and R. L. Byer, *J. Chem. Phys.* **65**, 5530 (1976).
26. V. I. Fabelinsky, B. B. Krynetsky, L. A. Kulevsky, V. A. Mishin, A. M. Prokhorov, A. D. Savel'ev, and V. V. Smirnov, *Opt. Commun.* **20**, 389 (1977).
27. A. Owyoung, C. W. Patterson, and R. S. McDowell, *Chem. Phys. Lett.* **59**, 156 (1978).
28. H. Lotem, R. T. Lynch, Jr., and N. Bloembergen, *Phys. Rev. A* **14**, 1748 (1976).
29. A. C. Eckbreth; *Appl. Phys. Lett.* **32**, (1978).
30. M. D. Levenson, *IEEE J. Quant. Electron.* **QE-10**, 110 (1974).
31. G. R. Meredith, R. M. Hochstrasser, and H. P. Trommsdorff, in "Advances in Laser Chemistry" (A. H. Zewail, ed.) (Springer Series in Chemical Physics 3). Springer-Verlag, Berlin, 1978.
32. L. A. Carreira, L. P. Goss, and T. B. Malloy, Jr., *J. Chem. Phys.* **69**, 855 (1978); **66**, 4360 (1977).
33. J. P. Coffinet and F. de Martini, *Phys. Rev. Lett.* **22**, 60, 752 (1969); in "Nonlinear Spectroscopy" (N. Bloembergen, ed.), pp. 319–349. North-Holland Publ., Amsterdam, 1977.
34. F. de Martini, G. Giuliani, P. Mataloni, E. Palange, and Y. R. Shen, *Phys. Rev. Lett.* **37**, 440 (1976).
35. A. Laubereau and W. Kaiser, *Rev. Mod. Phys.* **50**, 607 (1978), C. H. Lee and D. Ricard, *Appl. Phys. Lett.* **32**, 168 (1978).
36. J. P. Heritage, *Appl. Phys. Lett.* **34**, 470 (1979).
37. W. M. Tolles and R. D. Turner, *Appl. Spectrosc.* **31**, 96 (1977).

Chapter 5
MULTIPHOTON ABSORPTION

5.1 INTRODUCTION

The term multiphoton absorption has been used to refer to a variety of disparate effects which share only the one property of requiring the destruction of two or more quanta of electromagnetic radiation somewhere in the process [1–9]. The simplest phenomenon of this sort might be called a "multiphoton transition via virtual intermediate states" and is diagrammed in Fig. 5.1a. A quantum system initially in state $|g\rangle$ absorbs simultaneously N quanta of radiation in a transition to state $|t\rangle$. The radiation quanta can have different energies and wave vectors, but no sum of such quanta except the sum needed to reach state $|t\rangle$ is equal to an energy level separation in the quantum system. The dynamics of such a transition is correctly described by the two-level model of Chapter 2 with the effective Hamiltonian in (2.7.8). Alternatively, that Hamiltonian can be inserted in Fermi's golden rule number two,

$$\Gamma_{tg} = (2\pi/\hbar)|\langle t|\mathscr{H}_1|g\rangle|^2 \rho\left(\sum_\omega \hbar\omega\right) \qquad (5.1.1)$$

to evaluate the steady-state transition rate. In (5.1.1) $\rho(\sum \hbar\omega)$ is the density of states at the final energy, and for a sample of $\mathscr{N}$ atoms, the total transition rate is $\mathscr{N}$ times larger than the rate in (5.1.1).

At the opposite side of the spectrum of multiquantum effects stands the "stepwise" transition in (5.1b) where each photon is resonant with a single dipole-allowed transition. If the fields spanning each successive energy

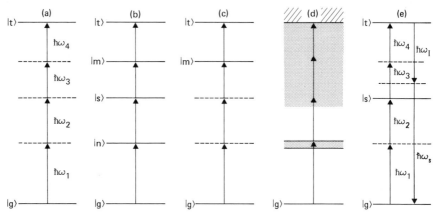

Fig. 5.1 Various multiphoton absorption processes. (a) A four-photon absorption process with no intermediate resonances; (b) a "stepwise" transition. States $|g\rangle$, $|s\rangle$, and $|t\rangle$ have the same parity, while state $|m\rangle$ and $|n\rangle$ have the opposite parity. (c) A three-photon resonant four-photon absorption; (d) a multiphoton dissociation process involving closely spaced intermediate states forming a "quasi-continuum." (e) A doubly resonant six-wave mixing process involving two- and four-photon resonances.

gap are active during successive time intervals, the process can be decoupled into two-level single photon transitions cascaded one after another. The methods of Chapter 2 can then be applied to each two-level transition in turn.

In the middle of the spectrum fall the "M-photon resonant N-quantum transitions," diagrammed schematically in Fig. 5.1c. If there is only one such intermediate resonance, it is possible to employ one of the exact solutions of the three-level system along with interaction operators for $|g\rangle|m\rangle$, $|m\rangle|t\rangle$, and $|g\rangle|t\rangle$ approximated by the techniques in (2.7). Needless to say, this is a more involved calculational program. If the final state $|t\rangle$ is a continuum and if the amplitude for reaching it without encountering state $|m\rangle$ is small, one can simplify the problem by considering the dynamics of the $|g\rangle$, $|m\rangle$ system in detail, but treating the $|m\rangle \to |t\rangle$ transition in a rate equation approximation.

When more than one such intermediate level is near resonance, the range of possible phenomena becomes too extensive to deal with except by numerical methods or gross approximation [10]. Some possibility of qualitative understanding remains when the coupling between one pair of levels can be shown to be much stronger than the others, or when one state broadens into a structureless continuum [11].

In molecular systems, there can be so many near-resonant levels that they form a "quasi-continuum" as in Fig. 5.1d. Successively and simultaneously

5.1 Introduction

absorbed quanta raise the average excitation level of such a system, but the overall phenomenon is probably better discussed as thermodynamics than as quantum electronics [12].

In most experiments, the final state $|t\rangle$ has some property which facilitates detection of population in that level. The final state may fluoresce at a distinct wavelength, be easily ionized by static fields, react chemically, ionize spontaneously, or dissociate. In a few techniques, resonances in a nonlinear susceptibility are detected by collecting the coherent radiation generated as the result of a nonlinear mixing process [13]. In such a case, phase matching can become a serious issue, and occasionally the information on oscillator strengths, etc. extracted from phase matching curves is the goal of the experiment. Such a coherent mixing process is indicated diagrammatically in Fig. 5.1e.

The nonlinear susceptibility plays an important role in coherent mixing techniques used to study multiquantum absorption. It is also true that the multiphoton transition rate itself is proportional to the imaginary part of a nonlinear susceptibility tensor. Many multiquantum experiments are, however, performed under transient conditions in which the steady-state assumptions underlying the derivation of the nonlinear susceptibility are not fulfilled.

The high-intensity lasers that must be employed in experiments on high-order nonlinearities often cannot be modeled accurately as coherent, transform limited pulses. A great deal of effort has been expended in trying to understand the consequences of finite laser bandwidth, etc. While this subject remains controversial, it seems that some qualitative understanding can be obtained by treating the dephasing of the radiation as a transverse relaxation process contributing to T_2^{-1} [14,15]. Such will be the approach taken here, it has been "justified" elsewhere. In this approximation, the onset of the steady-state regime occurs more quickly than would be predicted from a consideration of the isolated quantum system, and thus, Eq. (5.1.1) and the nonlinear susceptibility can be used widely.

Multiquantum processes are important spectroscopically for their ability to probe high-lying levels and levels forbidden for linear absorption spectroscopy. Inhomogeneous broadening can be eliminated in certain geometrical configurations, facilitating accurate assignment of term values. Isotope selectivity has been demonstrated for some efficient ionization and dissociation processes, stimulating hope of technological applications. Other processes are sensitive enough to detect single atoms of one species in a gas of another species at standard temperature and pressure. And some nonlinear processes probe tantalizing intellectual puzzles. For specialists, there is an extensive literature, for novices, the rest of this chapter [1-15].

5.2 DOPPLER-FREE TWO- AND THREE-PHOTON ABSORPTION

In the rest frame of an atom moving with velocity **v**, a light wave that has the frequency ω and wave vector **k** in the laboratory frame appears Doppler shifted to frequency $(\omega + \mathbf{k} \cdot \mathbf{v})$. Atoms with a transition frequency $\Omega = (E_t - E_g)/\hbar$ would be resonant with the incident light only if the atomic velocity were within a band centered at $v_z = (\omega - \Omega)c/\omega$. In a vapor where the atoms have a range of velocities, some atoms are always resonant, and the width of the absorption feature reflects the width of the velocity distribution. This is the inhomogeneous broadening removed by the saturation techniques of Chapter 3.

In the case where two laser beams propagate in opposite directions through a vapor as in Fig. 5.2 and the wave vectors are $\mathbf{k}_1 = \mathbf{k}$ and $\mathbf{k}_2 = -\mathbf{k}$ at every point, the motion of the atom shifts the frequency of one beam to $\omega + \mathbf{k} \cdot \mathbf{v}$ while the counterpropagating wave is shifted to $\omega - \mathbf{k} \cdot \mathbf{v}$. The frequency

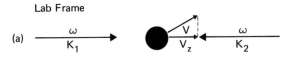

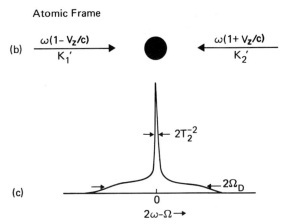

Fig. 5.2 The essence of Doppler-free two-photon absorption. (a) The two counterpropagating waves with frequency ω and the moving atom are illustrated at center. The sum of the two frequencies is constant in the reference frame of the moving atom in (b). (c) The expected spectrum with Doppler-free peak superimposed on a Doppler broadened background. The area of the narrower peak is twice that of the broad background.

5.2 Doppler-Free Two- and Three-Photon Absorption

of the sum, however, remains 2ω. Atoms with a photon transition at $2\omega = \Omega_{tg} = (E_t - E_g)/\hbar$ would be excited whatever their velocities. The transition would be homogeneously broadened, with a width given by the transverse decay rate and much, much narrower than the corresponding Doppler broadened transition. Since every atom participates in the absorption process, this nonlinear effect can be as strong as the one-photon resonant saturation nonlinearities discussed in Chapter 3.

The collapse of the Doppler width illustrated in Fig. 5.2 is an instance of a more general phenomenon possible for all multiphoton transitions. If M-photons with wave vectors $\mathbf{k}_i$ are absorbed and P-photons with wave vectors $\mathbf{k}_j$ are emitted in the course of an $N = M + P$ photon transition, the Doppler width would be

$$\Omega_D = \left| \sum_{i=1}^{M} \mathbf{k}_i - \sum_{j=1}^{P} \mathbf{k}_j \right| v_0, \tag{5.2.1}$$

where v_0 is the thermal velocity [1]. Clearly $\Omega_D = 0$ when $\sum_{i=1}^{M} \mathbf{k}_i = \sum_{j=1}^{P} \mathbf{k}_j$ which corresponds to the wave vectors forming a closed figure. Several such situations appear in Fig. 5.3. The rates of such transitions can be represented formally by Eq. (5.1.1) with the density of states factor

$$\rho = \left\{ \pi T_2 \left[\left(\sum_{i=1}^{M} \omega_i - \sum_{j=1}^{P} \omega_j - \hbar^{-1}(E_t - E_g) \right)^2 + T_2^{-2} \right] \right\}^{-1}, \tag{5.2.2}$$

where ω_i and ω_j correspond to absorbed and emitted photon frequencies, respectively.

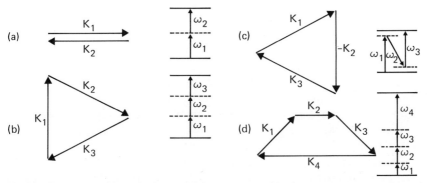

Fig. 5.3 Four types of Doppler-free multiphoton process. (a) The wave-vector summing and level diagrams for Doppler-free two-photon absorption. (b) Three-photon absorption and (c) a Cagnac-type interaction with two-photons absorbed and one stimulated emission. (d) A Doppler-free four-photon absorption with widely different frequencies. In each case, the wave vectors of the interacting beams sum to zero.

This kinetic analysis assumes that the detuning of any intermediate states from the sum of photon energies is much greater than the Doppler width of the corresponding transition. In such a case, the effective operators of (2.7) can be used without terms proportional to velocity in the denominators.

When all of the beams have the same frequency, the Doppler-free multi-photon process is usually accompanied by several Doppler broadened multiquantum processes each of which corresponds to an interaction with N quanta from a single beam. For the Doppler-free two-photon absorption process in which beams of amplitude $\mathbf{E}_+$ and $\mathbf{E}_-$ propagate with wave vectors $\mathbf{k}$ and $-\mathbf{k}$ and frequency ω through a medium where the Doppler width Ω_D is much greater than T_2^{-1}, the total transition rate per atom is

$$\Gamma_{tg} = \left\{ \frac{1}{16\hbar^4} \left| \sum_n \frac{(\boldsymbol{\mu}_{tn} \cdot \mathbf{E}_+)(\boldsymbol{\mu}_{ng} \cdot \mathbf{E}_-) + (\boldsymbol{\mu}_{tn} \cdot \mathbf{E}_-)(\boldsymbol{\mu}_{ng} \cdot \mathbf{E}_+)}{\Omega_{ng} - \omega} \right|^2 \right.$$

$$\times \frac{T_2^{-1}/\pi}{(\Omega_{tg} - 2\omega)^2 + T_2^{-2}} + \left[\left| \sum_n \frac{(\boldsymbol{\mu}_{tn} \cdot \mathbf{E}_+)(\boldsymbol{\mu}_{ng} \cdot \mathbf{E}_+)}{\Omega_{ng} - \omega} \right|^2 \right.$$

$$\left. \left. + \left| \sum_n \frac{(\boldsymbol{\mu}_{tn} \cdot \mathbf{E}_-)(\boldsymbol{\mu}_{ng} \cdot \mathbf{E}_-)}{\Omega_{ng} - \omega} \right|^2 \right] \frac{\exp[-(2\omega - \Omega_{tg})^2/\Omega_D^2]}{2\sqrt{\pi}\Omega_D} \right\}. \quad (5.2.3)$$

This line-shape function is illustrated at the bottom of Fig. 5.2. If the electric field amplitudes are equal, the area of the narrow Doppler-free peak implied by the first term is twice that of the Doppler-broadened component. Analogous effects occur for higher-order processes involving several beams at one or more distinct frequencies.

The angular momentum selection rules for multiquantum transitions can be obtained by sequentially applying the selection rules for the single-photon transitions occurring virtually in the overall process, and summing the resulting amplitudes with the correct weighting factors. For an Nth-order dipole process, the change in total angular momentum must be $\leq N$, whereas the emission or absorption of each photon alters the angular momentum by 0 or ± 1 depending upon the polarization. Simplifications occur when the fine and hyperfine interactions split the virtual intermediate states by an amount less than the detuning. The summation over all substates of the virtual intermediate multiplets then leads to a scalar operation and the selection rules depend only upon the character of the initial and final states. The electromagnetic interaction couples only to the orbital angular momentum, and the intensity ratios for various components can be obtained using Clebsch–Gordon coefficients or 6-J symbols [2, 5, 16].

Angular-momentum selection rules can be used to eliminate the Doppler-broadened background in an even-quantum absorption process when the initial and final state angular momenta are equal. The initial demonstration

5.2 Doppler-Free Two- and Three-Photon Absorption

of this background suppression involved a two-photon transition between atomic S states. Since the orbital angular momentum vanishes in the initial and final states, the selection rule $\Delta L = \Delta m_L = 0$ applies. Absorption of two quanta from a single circularly polarized beam requires $\Delta m_L = \pm 2$, clearly impossible. Transitions can occur only for circularly polarized light when the atoms absorb one quantum with angular momentum $+1$ from one beam and a quantum with angular momentum -1 from the oppositely propagating beam. Thus, the transition is allowed only when the laser beams are circularly polarized with the photon spins pointing in opposite directions along a fixed axis, or—in common optical terminology—where the two counter-propagating beams have the same sense of circular polarization [2].

The optical Stark effects referred to in Eq. (2.7.10) scale linearly with intensity

$$\Delta E_u = \frac{1}{4\hbar} \sum_n \left\{ \frac{|\mathbf{\mu}_{un} \cdot \mathbf{E}|^2}{\Omega_{nu} - \omega} + \frac{|\mathbf{\mu}_{un} \cdot \mathbf{E}|^2}{\Omega_{nu} + \omega} \right\}. \qquad (5.2.4)$$

Since the two-photon transition rate scales as the square of the intensity and since higher-order multiquantum processes show an even more pronounced intensity dependence, it is virtually impossible to perform high-resolution multiquantum absorption experiments without noticeably shifting the levels. When the laser intensities are nonuniform spatially, atoms in different regions experience different shifts. When pulsed lasers are employed, the shifts become time dependent. In the context of high-resolution spectroscopy, both these effects cause broadening, and none have been included explicitly in (5.2.3). Also causing broadening are excited-state absorption processes, particularly ionization of the final state reached by the Doppler-free multiquantum process. Of course, ordinary two-level saturation can contribute additional width when the coupling Hamiltonian is strong enough. To some extent, all of these broadening and shifting effects can be reduced along with the transit time broadening that results from the finite diameter by the Ramsey-fringe technique discussed in Chapter 6.

The earliest Doppler-free two-photon experiments were performed in atomic sodium which has two strong transitions within the gain band of rhodamine 6G. Each of the four early investigators detected fluorescence resulting from the decay of the state excited by the two-photon process [2]. The hyperfine splitting of the 3^2S–5^2S transition at 6022 Å as resolved by one of the first pulsed laser experiments is shown in Fig. 5.4. The suppression of the Doppler background in the correct circular polarization condition is clearly evident. The width of the Doppler-free components is equal to the laser linewidth. An intermediate-field Zeeman spectrum taken on this transition with a cw laser [17] is shown in Fig. 5.5. The positions and strengths of the individual components can be calculated considering only the Zeeman

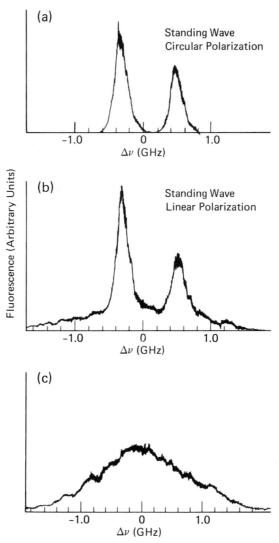

Fig. 5.4 Two-photon absorption signals on the 3S → 5S two-photon transition in atomic sodium. The experimental traces plot the observed fluorescence intensity at 330 nm as a function of laser frequency. (c) The Doppler broadened line shape found with traveling-wave excitation. (b) The Doppler-free resonances superimposed on the Doppler-broadened background. With circular polarization, background is suppressed leaving the lineshape shown in (a). The splitting between the lines reflects the difference in hyperfine constants between the upper and lower levels (from Ref. [2]).

5.2 Doppler-Free Two- and Three-Photon Absorption

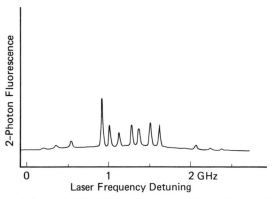

Fig. 5.5 Zeeman splitting of the 3S → 5S two-photon transition at intermediate fields (800 G). The electron and nuclear spins are beginning to decouple and 13 of the 14 transitions are nondegenerate. The use of a cw laser rather than a pulsed source results in substantially improved resolution (from Ref. [17]).

structure of the initial and final states. The effect of tuning near the 3^2S–$3^2P_{3/2}$, $3^2P_{1/2}$ resonances using oppositely propagating beams with slightly different frequencies is shown in Fig. 5.6. Not only is the two-photon cross section for populating the $4^2D_{5/2}$ and $4^2D_{3/2}$ states enhanced, but the $D_{5/2}$ cross section is enhanced only by the $P_{3/2}$ resonance, a result predictable from the single-photon selection rules. The cross section for the $D_{3/2}$ state shows a resonance at each of the intermediate levels and a minimum due to quantum mechanical interference between the two. The resonant character of the optical Stark shift was also demonstrated using this transition and lasers of differing frequency [1,2].

The first demonstration of a Doppler-free three-photon interaction also exploited the strong resonances in sodium, but in this case the resonance was the $3^2S_{1/2}$–$3^2P_{1/2}$ component of the D lines. The level diagram and wave-vector sum figure for the three-photon process appears in Fig. 5.3. Two quanta of frequency ω_1 are absorbed while one of frequency ω_2 is emitted. Both frequencies are very near the frequency of the one-photon allowed transition, and thus there is considerable resonant enhancement. A triple monochromator is necessary to separate the fluorescence due to the decay of the $3^2P_{1/2}$ state from scattered laser radiation. The four hyperfine components of the three-photon transitions were resolved with cw lasers and an instrumental resolution of 60 MHz [1].

Doppler-free two-photon spectroscopy has by now been employed to study a great many transitions between states of the same symmetry and to determine term values, line widths, pressure shift, and broadening coefficients and other parameters. Most of the effort has focused on atoms including

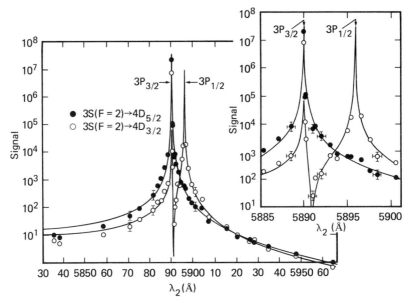

Fig. 5.6 The variation of the Doppler-free two-photon cross section for the 3S → 4D transitions in the region of the $3P_{1/2}$ and $3P_{3/2}$ intermediate resonances. Two lasers with slightly different frequencies are required to access these resonances and thus some residual Doppler width remains. A destructive interference between resonant amplitudes is clearly visible for the $3S$–$4D_{3/2}$ cross section (from Ref. [2]).

hydrogen, the alkalis, helium, neon, and thallium, but there has also been work on vibrational transitions in such molecules as NH_3 and CH_3F, and rovibronic transitions in Na_2, [1, 18].

The hydrogen program of Hänsch deserves special emphasis because of its potential for fundamentally improving our knowledge. The two-photon transition between the $1\,^2S$ and $2\,^2S$ states requires a wavelength of 2430 Å, exactly one-half the wavelength of the blue Balmer line. By generating the shorter wavelength by second harmonic generation and by comparing the frequency of the two-photon transition to that of a particular component of the one-photon line, the Lamb shift of the $1\,^2S$ state can be measured (see Fig. 3.21). Moreover, the decay rate of the $2\,^2S$ state is 7 s^{-1} under ideal conditions. Measuring the term value to such accuracy would greatly improve the precision of the Rydberg constant and possibly allow the meter and second to be defined in terms of this fundamental transition. Precision measurement of the hydrogen–deuterium isotope shift would improve the accuracy with which the relative masses of the proton, neutron, and electron are known [19].

Another remarkable effort focused on measuring the term values for Rydberg states of rubidium by two-photon absorption. Rydberg atoms

5.2 Doppler-Free Two- and Three-Photon Absorption

with high principal quantum numbers ($n \sim 70$) have radii approaching 1 μm and are remarkably delicate. Lee et al. excited these states by two-photon absorption using a cw laser and detected them by field ionization in a special field-free thermonic diode [20]. All the observed term values could be related to a quantum defect formula

$$E_n = P_1 - [R_m/(n - \mu_n)^2] \tag{5.2.5}$$

with an accuracy of less than 10^{-3} cm^{-1} for $n > 15$. The quantum defect varied with effective quantum number as

$$\mu_n = P_2 + P_3/n_0^{*2} + P_4/n_0^{*4} \quad \text{and} \quad n_0^* = n - P_2. \tag{5.2.6}$$

Thus, the term values can be employed as a secondary standard for frequency calibration. For ^{85}Rb, the constants are $P_1 = 33{,}690.7989(2)$ cm^{-1}, $P_2 = 3.13109(2)$, $P_3 = 0.204(8)$, $P_4 = -1.8(6)$, and $R_m = 109736.615$ cm^{-1}.

The collisional properties of these Rydberg atoms remain a bit of a puzzle. Plots of the pressure broadening for rubidium versus principal quantum number [21] are shown in Fig. 5.7. Since the time of Fermi, it has been

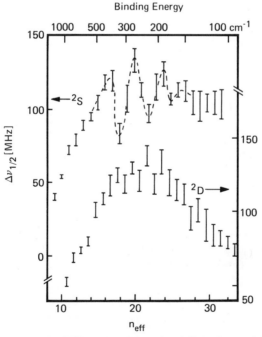

Fig. 5.7 Observed linewidths (FWHM) of the $5S \to nS$ and $5S \to nD$ two-photon transitions in Rb at 60-mTorr pressure. The scale on the left is for S levels, and that on the right is for the D levels. The oscillatory dependence on the effective quantum number $n_{\text{eff}} = n_0 - \mu_n$ remains a mystery (from Ref. [21]).

expected that the initial increase of these parameters would terminate and the shift and broadening either decrease monotonically or remain constant for very high n. The oscillatory behavior in the transition region was unexpected, and—at present—unexplained.

5.3 MULTIQUANTUM IONIZATION

5.3.1 Nonresonant Ionization

When a powerful laser beam is focused to a small spot in air or another gas, the intense optical fields can ionize the gas causing a visible spark. This effect led to the realization that a sufficiently powerful laser could ionize any species by a nonresonant multiquantum process no matter what the photon energy. The number of quanta required—the so-called order of the interaction N—is the first integer larger than $E_I/\hbar\omega$ where E_I is the ionization limit of the species [5].

Presently, the process leading to the visible spark is understood as an avalanche ionization process similar to more familiar dielectric breakdown effects and not closely related to the multiquantum ionization discussed here. Still, Maker's 1964 observation sparked a great deal of theoretical and experimental effort. Since lasers powerful enough to cause multiquantum ionization were not tunable until recently, most experiments were conceived as nonresonant, that is, involving the ground state, the ionization continuum, and only virtual states in between.

The transition rate can be calculated from (5.1.1) with the interaction Hamiltonian evaluated to Nth order and the correct continuum density of states substituted for ρ. It has become conventional, however, to write the transition probability for N-photon ionization in terms of a generalized total cross section

$$\Gamma_{fg}^{(N)} = \hat{\sigma}_N F^N, \qquad (5.3.1)$$

where F is an effective photon flux (in photons/cm^2 s) which is equal to $I/\hbar\omega$ for the perfectly coherent fields represented here as classical waves. The generalized total cross section $\hat{\sigma}_N$ has dimensions cm^{2N} s^{N-1} and is given by

$$\hat{\sigma}_N = \left(\frac{2\pi\omega}{\hbar}\right)^N \frac{m|\mathbf{K}|}{4\pi^2\hbar} \int |\langle f|\mathcal{H}_1|g\rangle|^2 \, d\Omega_k \qquad (5.3.2)$$

where m is the electron mass, $\mathbf{K}$ the wave vector of the outgoing photoelectron which is related to the photon and ionization energies by

$$N\hbar\omega - E_I = \hbar^2|\mathbf{K}|^2/2m, \qquad (5.3.3)$$

5.3 Multiquantum Ionization

and where the electric field amplitudes in $\mathcal{H}_I$ have been replaced by complex polarization vectors of unit magnitude. The integration is over all directions of propagation for the emitted electron. The final continuum state can be written as

$$|f(\mathbf{r})\rangle = 4\pi \sum_{L=0}^{\infty} i^L e^{-i\delta_L} G_L(K,r) \sum_{M=-L}^{+L} Y_L^{*M}(\Theta,\Phi) Y_L^M(\theta,\phi), \quad (5.3.4)$$

where δ_L is the phase shift and G_L the radial part of the partial wave of angular momentum L, and Y_L^M the spherical harmonics. In spherical coordinates the $\mathbf{r}$ and $\mathbf{K}$ vectors are represented as (r,θ,ϕ) and (K,Θ,Φ), respectively. Calculation of this generalized cross section requires summation over all atomic states, including continuum states, and is a considerable exercise in atomic physics. Results of a typical calculation are shown in Fig. 5.8; in the regions near intermediate state resonances, the value of the cross section lies above the top of the figure [5].

The experimental measurement of nonresonant multiquantum ionization cross sections has proved remarkably difficult. The signals are quite small and easily obscured by backgrounds from the resonant ionization of contaminant species. Quantitative measurement of the photon flux at focus is necessary for each shot, as is near perfect control of the laser parameters.

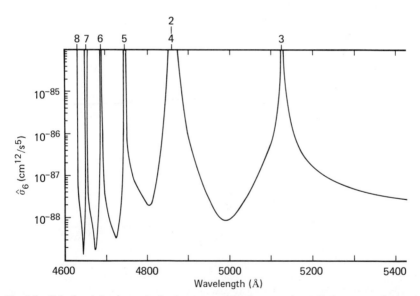

Fig. 5.8 Calculated six-photon ionization cross section for ground state hydrogen as calculated by Karule. The numbers at top indicate the quantum numbers of intermediate resonances. While spin-orbit splitting is unimportant in hydrogen, this cross section shows the qualitative features expected for more complex atoms (from Ref. [5]).

Finally, since the Coulomb potential supports an infinite number of bound states the energies of which are sensitive to external perturbations, it is difficult to ensure a true nonresonant condition, especially for high-order processes.

The experimental strategy has been to collect positively charged ions rather than electrons in order to reduce background levels, and to do mass selection in an attempt to verify the species. Atomic beams and loosely focused laser beams have proved valuable in defining the focal volume. Even so, detailed calculations are necessary to relate the observed ionization signal to the theoretical rate at every point in the focal volume [4].

It is crucial to control the spectrum and thus the photon correlation properties of the laser source. The average of the Nth power of the photon flux is larger by a factor of $N!$ for a chaotic field—produced, for example, by amplified spontaneous emission—than for a coherent field with the same measured intensity [15]. An experiment to measure a multiphoton cross section is considered successful if it comes within two orders of magnitude of the theoretical prediction.

The difficulties in measuring the absolute cross section have focused efforts on measurements of relative cross sections for light of different polarizations. In Fig. 5.9 are shown the different angular momentum channels open for multiphoton ionization with linear and circularly polarized light for atoms with and without spin-orbit coupling. Clearly the angular momentum selection rules for dipole transitions allow more open channels for linear than for circularly polarized light. The ionization cross section depends both on the number of channels and the magnitude and phase of the matrix elements. In most experiments to date, the cross section for circularly polarized light has exceeded that for linearly polarized light. These ratios provide some information on the ratios of bound-free matrix elements [5, 22].

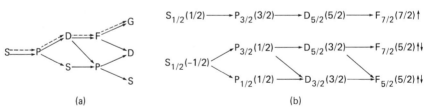

Fig. 5.9 Angular momentum channels available for four-photon ionization of an S state. (a) The linearly (—) and circularly (---) polarized cases are illustrated when spin-orbit coupling can be ignored. (b) The case for atoms with spin-orbit split levels and right circular polarization. The numbers in parentheses are the spin projection quantum numbers while the arrows illustrate the possible orientations of the free electron spin. Each channel corresponds to a group of levels with the same angular momentum quantum numbers but different energies that can appear as intermediate resonances (from Ref. [5]).

5.3 Multiquantum Ionization

The experimental order of the nonlinearity,

$$N_{\text{exp}} = \frac{\partial \log(\text{number of ions})}{\partial \log(\text{intensity})}, \qquad (5.3.5)$$

has been more successfully measured. At low intensities, N_{exp} has been shown to equal N for well-characterized systems. At high intensities, deviations result either from the "volume" or "saturation" effect (in which all the atoms in the region of maximum intensity are ionized while an increasing number at the edges of the beam continue to contribute signal) or from near-resonant effects. For a Gaussian beam focused by a spherical lens, the volume effect predicts a limiting $N_{\text{exp}} = \frac{3}{2}$ at high intensities. Ionization signal data from the noble gases obtained using a very powerful mode locked laser, mostly in the nonresonant regime [23] is shown in Fig. 5.10. Similar results have been obtained for the alkalis which have lower ionization potentials and hence lower-order nonlinearities [24].

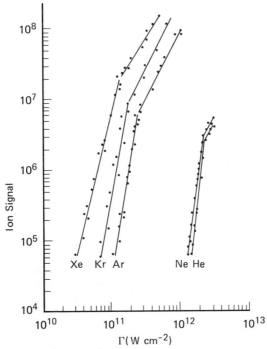

Fig. 5.10 Multiphoton ionization signals for the noble gases as a function of light intensity. The slope on such log–log plots is usually interpreted as the experimental index N_{exp}. At low power, these indices equal the number of quanta required for ionization by the 1.06-μm radiation. At higher intensities, an intermediate level is Stark shifted into resonance and the index decreases (from Ref. [23]).

5.3.2 Singly Resonant Multiquantum Ionization

When the energy of M radiation quanta approaches the energy of a bound state $|t\rangle$, the cross section for N quantum ionization greatly increases as the result of resonant effects. The terms in (5.1.1) that involve this resonant level dominate the ionization process, and the dynamics of the resonant intermediate transition becomes significant. Since the final state of the N-photon process is still a continuum state, the resulting three-level system is not as complex as it might have been. The ionization processes depopulating the resonant state $|t\rangle$ can be treated in the rate equation approximation with a rate given by (5.1.1) with the state $|t\rangle$ substituted for $|g\rangle$ and the effective order $N - M$ substituted for N. The dynamics of the two-level system $|g\rangle - |t\rangle$ coupled by the M-photon transition can then be dealt with using the techniques of Chapter 2. Because of the enormous intensities occurring in such experiments, all the effects of coherent dynamics and level shifts should be included in the description.

In addition to an increased cross section, the experimental signatures of an M-photon resonant N-quantum ionization include a deviation of the experimental order $N_{\exp}$ from N. There appear to be three cases:

(1) When the coupling between the resonant intermediate state $|t\rangle$ and the continuum is stronger than that between $|t\rangle$ and $|g\rangle$, an intensity level is reached where every atom reaching $|t\rangle$ is ionized. The experimental index $N_{\exp} = M$ under these circumstances, commonly occurring when the laser linewidth is broader than the atomic linewidths.

(2) When the coupling between the ground and resonant intermediate state is much stronger than the coupling between $|t\rangle$ and the continuum $|f\rangle$, half the atoms are in state $|t\rangle$ on the average, no matter what the intensity. The experimental index $N_{\exp} = N - M$ reflects the scaling law of the unsaturated ionization process.

(3) When the optical Stark shifts act to move the resonant intermediate level *closer* to the energy of M photons, the experimental index can be larger than N. When the shifts act to push the level *further* from resonance, $N_{\exp}$ drops below N. There thus can be a characteristic dispersion shaped resonance in $N_{\exp}$ centered somewhat near the energy of state $|t\rangle$ *as shifted by* the average field in the experiment.

All three effects can occur simultaneously, and the resonant process can also be strong enough to ionize all the atoms in the focal volume, causing the signal to become independent of intensity. When these anomalies occur, the ratio of the cross sections for circular and linear polarization deviates from its low intensity value and becomes intensity dependent [25].

5.3 Multiquantum Ionization

Some qualitative understanding of the dynamics of these transitions can be obtained from an "extended two-level" model of Eberly [26]. In this model, the transition rate between ground and resonant intermediate state is given by

$$\Gamma_{tg}^{(M)}(\Omega_{tg} - M\omega) = \frac{\chi^2}{8} \frac{\Gamma_t}{(\Omega_{tg} - M\omega + sF)^2 + \Gamma_t^2}, \quad (5.3.6)$$

where χ^2 is the square of the M-photon Rabi frequency and is proportional to F^M. The detuning from resonance depends upon intensity through the linear Stark shift term

$$sF = \mathcal{H}_{1tt} - \mathcal{H}_{1gg}, \quad (5.3.7)$$

where diagonal matrix elements in (2.7.8) or the shifts in (5.2.4) can be used for a coherent field. For chaotic radiation, the Stark shift could be as much as a factor of 3 larger. The ionization rate of the resonant intermediate state is described by a nonresonant multiphoton cross section

$$\Gamma_{ft}^{(N-M)} = \hat{\sigma}_{(N-M)} F^{N-M}. \quad (5.3.8)$$

The effective transverse and longitudinal decay rates for the $|g\rangle - |t\rangle$ system include the effects of ionization and the finite laser bandwidth γ_L,

$$\Gamma_t = T_2^{-1} + \gamma_L + \tfrac{1}{2}\Gamma_{ft}^{(N-M)}, \qquad \Gamma_l = T_b^{-1} + \Gamma_{ft}^{(N-M)}. \quad (5.3.9)$$

The total N-photon ionization rate becomes

$$\Gamma^{(N)} = \tfrac{1}{8}\chi^2 \sigma_{(N-M)} F^{N-M} \frac{\Gamma_t/\Gamma_l}{(\Omega_{tg} - M\omega + sF)^2 + \Gamma_t^2 + \chi^2\Gamma_t/\Gamma_l}, \quad (5.3.10)$$

where the low intensity cross section can be identified as

$$\sigma_N F^N = \frac{\chi^2 \sigma_{N-M} F^{N-M}(T_2^{-1} + \gamma_L)T_b/8}{(\omega_{tg} - M\omega)^2 + (T_2^{-1} + \gamma_L)^2}. \quad (5.3.11)$$

The effective nonlinear index can be defined as

$$N_{\text{eff}} = \partial \log \Gamma^{(N)} / \partial \log F \quad (5.3.12)$$

and this quantity has the expected behavior in the simple limiting cases: at low intensity $N_{\text{eff}} = N$. When χ is the largest frequency and the two-level system saturates, $N_{\text{eff}} = N - M$, etc. It also predicts $N_{\text{eff}} > N$ when $sF > \sigma_{N-M}F^{N-M}$ and the Stark shift acts to bring the intermediate state near resonance. When the frequency is adjusted to achieve resonance in the presence of Stark shifts, Eq. (5.3.10) also reflects the variation of the ratio of cross sections for different polarization. Unhappily, the agreement between (5.3.10) [even when (5.3.10) augmented by numerical integration over space

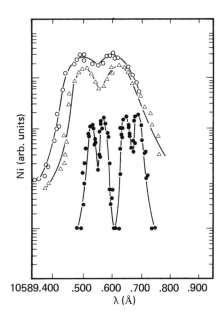

Fig. 5.11 Three-photon-resonant four-photon-ionization signals in cesium. At low intensities, all four components of the 6S → 6F three-photon transition are visible. At higher intensities, ionization broadening and Stark shifts are apparent (from Ref. [27]). $I = 2.2$ (●), 5.0 (△), and 7.3 (○) × 10^7 W cm^{-2}.

and time profiles] and the available data is not always quantitative [26]. Better accuracy has been obtained using numerical solutions of the master equations of Chapter 2 [27].

The most reliable experimental data have been obtained by the Service de Physique Atomique at CENS using a tunable single mode Nd:glass laser. Their data on four-photon ionization of cesium with three-photons resonant with the $6^2S - 6^2F$ transition [27] are shown in Fig. 5.11. The four-fine/hyperfine structure components are visible at the lowest intensity and the shift and broadening of the resonances are evident at higher intensity levels. The variation of the experimental order of the nonlinearity as a function of detuning from resonance along with a theoretical curve (which takes space and time inhomogenities of the intensity into account) is shown in Fig. 5.12.

When the exciting frequency is detuned from a particular intermediate state resonance, the amplitude for ionization via that resonance is greatly reduced. At some frequency, the resonant amplitude will have a magnitude equal to that of the sum of all other possible transitions, but opposite in phase. The result is a destructive interference similar to that appearing in Fig. 5.6. While these minima have been predicted by many theories (see Fig. 5.8), they have not yet been unequivocally observed in multiphoton ionization. The nonlinear cross section in the region of such a minimum is so small that otherwise negligible background effects overwhelm the experimental signals. Such effects include dimers which absorb radiation and

5.3 Multiquantum Ionization

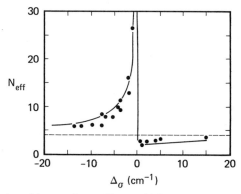

Fig. 5.12 The variation of the experimental index with laser detuning in cesium. The points are derived from data as in Fig. 5.11 while the solid line is a theoretical curve incorporating the effects of the space and time inhomogeneities of the laser beam (from Ref. [27]).

dissociate into ions, and fluorescence from the laser which can populate the resonant state. Still, quantum mechanics requires the presence of such destructive interference minima, and it should be possible to observe them in experiments similar to that which produced Fig. 5.11.

5.3.3 Multiple Resonance and Molecular Dissociation

Some of the complexities that can arise when there are multiple resonances between discrete states [28] are illustrated in Fig. 5.13. In this experiment, a strong field drives the $3\,^2S_{1/2}$–$3\,^2P_{1/2}$ transition of sodium while a weaker field couples the $3P_{1/2}$ level to the $4D_{3/2}$. Some of the atoms reaching the $4D_{3/2}$ state absorb an additional photon and ionize. The cross section for this process is very large because of the double resonant enhancement. The data plotted in Fig. 5.13 show the ionization signal as a function of the frequency of the field coupling the $3P_{1/2}$ and $4D_{3/2}$ states. Note that when the laser is tuned to the frequency predicted for the transition by conventional spectroscopic tables, the ion signal is at a minimum.

What has happened in this multiresonant process is that the coupling between the $3\,^2S_{1/2}$ and $3\,^2P_{1/2}$ states is strong enough to split the upper level into an Autler–Townes doublet [29]. The multiphoton ionization process can be resonant only when the energy of the second photon equals the energy difference between an Autler–Townes component and the $3D_{3/2}$ state. Additional structure could have appeared had the second field been strong enough. The ionization signal scales linearly with the intensity of the weaker beam, but does not depend upon the intensity of the stronger field in any simple way. Since the intensities in the active regions were not

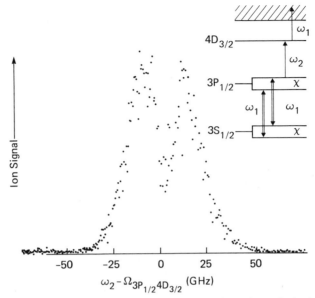

Fig. 5.13 Autler–Townes splitting of the multiply resonant three-photon ionization resonance in sodium. The inset shows the overall process in which a strong field resonant with the $3P_{1/2}$ level splits it into Autler–Townes components while a weaker beam tunes across the region of the $3P_{1/2}$–$4D_{3/2}$ resonance. A second photon from the strong beam causes ionization. Temporal and spatial nonuniformities broaden the resonances and reduce the splitting (from Ref. [28]).

constant in space or time, the Autler–Townes components in Fig. 5.13 are broader than would be predicted from the laser and atomic linewidths.

The enormous cross sections of the multiresonant processes compensate for their complexity. Polyatomic molecules such as SF_6, OsO_4, and CF_3Br can efficiently absorb 20 or more photons of 10-μm radiation, ultimately dissociating into an atom and a vibrationally excited fragment [8, 10, 12]. This process has drawn considerable attention because it can be isotopically selective; that is, all of the $S^{32}F_6$ can be made to dissociate into $SF_5 + F$ leaving the $S^{34}F_6$ unaffected. Only resonant vibrational transitions were thought to provide selectivity. The vibrational spectra, however, are quite complex, even for low-lying vibrations. Near the threshold for dissociation, the vibrational energy levels are thought to blend into a continuum. Because of the many resonant and near-resonant intermediate states, this multiphoton dissociation process can be efficient even at the intensity levels of a loosely focused TEA laser beam.

The experimental results indicate that the intensity does not really matter once it reaches a threshold of sorts. The dissociation efficiency and vibrational temperature depends instead on the energy fluence, the total energy

5.3 Multiquantum Ionization

passing through a unit area of the gas. In SF_6 it has also been shown that the energy deposited by the laser is not confined to a single mode of vibration, but distributed among modes on a picosecond time scale according to their statistical weights. Collisions with other molecules have nothing to do with the multiphoton dissociation process; the energy randomization is a feature of the internal dynamics of a polyatomic molecule. When sufficient energy has been absorbed, the probability for dissociation begins to rise. When the dissociation rate reaches 10^{-8} s^{-1} or so, the molecule falls apart and stops absorbing quanta, taking the lowest energy dissociation channel. The fragments, however, are vibrationally excited. For awhile, the molecule had contained more energy that equilibrium thermodynamics would allow [12].

The picture used to discuss this multiquantum dissociation process is illustrated in Fig. 5.14. The lower vibrational energy levels are nearly equally spaced, and a low-order ($N = 2, 3, \ldots$) multiquantum process populates higher vibrational states in a spectrally selective fashion. The rotational structure and the relatively large (10 cm^{-1}) Rabi frequencies weaken the resonance condition. These specifically excited levels decay rapidly into a statistical distribution of excitations. However, just above this intermediate region on the ladder of vibrational energy levels lies a "quasi-continuum" of levels. Every single quantum transition is resonant no matter what the initial state. Continuing laser irradiation populates higher and higher vibrational states. The dynamics of this process remains controversial; some

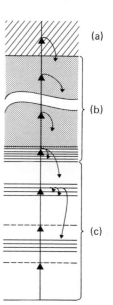

Fig. 5.14 Multiphoton dissociation in polyatomic molecules. The process is discussed in terms of (c) discrete multiphoton transitions among relatively low-lying levels followed by (b) automatically resonant transitions in a region where the levels are closely enough spaced to form a "quasi-continuum." (a) Eventually, the energy of the true dissociation continuum is reached, but additional quanta can still be absorbed for a short period of time. Decay processes are operative from all intermediate levels.

authors describe it as Joule heating, others by a set of rate equations for the resonant levels, still others by a coherent multiquantum process [10]. It is clear, however, that for intensities near 200 MW/cm^2, the process results in the absorption of 30 photons or so per molecule in 50 ns or so. The resulting population distribution includes a high-energy tail that extends beyond the dissociation limit; the molecules constituting that tail dissociate [12].

At least 30 equations would be required to describe this process in terms of Fermi golden-rule type rates, and the rates themselves would have to be guessed. A density matrix calculation would require 500 equations. While such treatments are not beyond the capabilities of those most interested in the problems, a description of the vibrational population distribution in terms of an effective temperature $T_{\text{eff}}(t) = \langle N(t) \rangle \hbar\omega / s k_B$—where $\langle N(t) \rangle$ is the average number of quanta absorbed per molecule at time t and s is the number of vibrational modes—may prove sufficient.

Ackerhalt and Eberly have pointed out a condition in which the coherence effects in multiply resonant multiquantum processes become unimportant enough that rate equations provide an adequate description of the dynamics [11]. In their condition, the multiquantum process must end in a continuum, such as the ionization or dissociation continuum. The coupling between the continuum and the discrete state just below it must be the strongest in the system, with the coupling between each lower pair of discrete levels becoming successively weaker and weaker. Under these conditions, the width of the highest discrete state is dominated by the dissociation or ionization rate. The Rabi frequency for the highest pair of discrete states is less than the ionization or dissociation width and thus coherence effects are relatively unimportant. The coupling between these two levels, however, broadens the lower level, sufficiently to make its width exceed the Rabi frequency for the next lower pair of levels, etc. In the Ackerhalt–Eberly condition, the resonant couplings themselves destroy all coherence. Unfortunately, the intensities required to achieve this condition, even on exact resonance, can be prohibitively large in practice. More common is the case in Fig. 5.13 where the lower transitions have the stronger couplings.

The most common way of avoiding coherence problems in multiquantum transition among discrete states is to use low-power broad-band lasers, and to hope that the laser linewidths are large enough and the intensities low enough that the effective relaxation rates will be larger than the Rabi frequencies. A more certain technique for preventing the propagation of coherence through a process is to turn off the laser coupling one transition on a multiquantum ladder before turning on the one between the next pair of levels. It is important, of course, to start at the bottom of the ladder. If the

5.3 Multiquantum Ionization

lowest level in such a stepwise process is labeled zero, the rate equation limit predicts that the population of the Nth level can be as large as $1/2^n$ of the initial population of the zeroth level. Using π pulses, much more population can be transported up the ladder.

A typical stepwise multiphoton ionization experiment is illustrated in Fig. 5.15 [30]. Three lasers interact with an atomic sample placed between

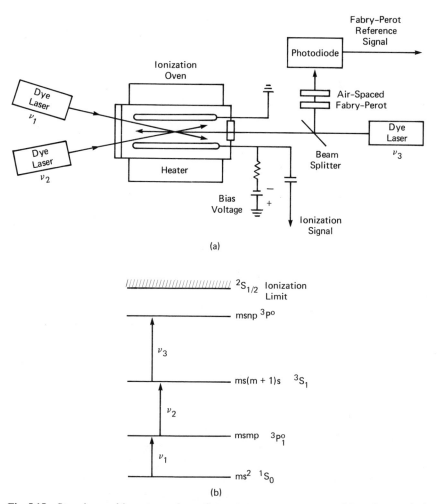

Fig. 5.15 Stepwise multiquantum absorption spectroscopy apparatus (a) and a typical multiresonant energy level scheme (b). The frequencies of two lasers are adjusted to populate a convenient intermediate level. A later tunable pulse then scans the spectral range of interest. Collisions and stray fields produce enough ionization to detect with a thermonic diode (from Ref. [30]).

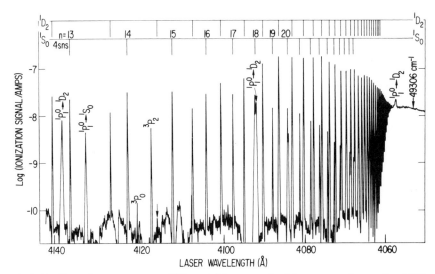

Fig. 5.16 Two-photon spectrum of $^1S_0$ and $^1D_2$ states of calcium converging on the 49306-cm^{-1} ionization limit. One-photon transitions are indicated. This trace is typical of multiquantum ionization data taken with apparatus like that in Fig. 5.15 (from Ref. [31]).

the plates of an ionization cell. The first two lasers are tuned to known transitions and are used to populate a relatively long-lived excited state. The frequency of the third laser is tuned to scan over energy levels being studied. These highly excited states ionize spontaneously or as the result of collisional or chemical processes and the resulting electrical signals are plotted as a function of frequency. Alternatively, a voltage pulse can be applied to the ionization plates causing field ionization of the highly excited levels. The ions produced can be directed into a secondary emission electron multiplier for amplification if the electrical signals are otherwise insufficient. A portion of the spectrum of calcium obtained in this fashion is shown in Fig. 5.16. The term values and oscillator strengths obtained in this fashion can be explained using multichannel quantum defect theory.

The Stark split components of some high Rydberg states of sodium studied by applying a weak voltage to the ionization plates during the laser pulse and gating the ion detection electronics are shown in Fig. 5.17. The spectra are plotted as a function of the Stark field. The resonances vanish when that field becomes strong enough to cause field ionization before the electronics is activated, and reappear when the gate time is set to zero [7].

These stepwise multiquantum techniques allow the study of very highly excited levels using visible and near-ultraviolet lasers. States can be studied which are not coupled strongly enough to the ground state to produce detectable absorption. Thus, even-parity Rydberg series of the alkalis and

5.4 Nonlinear Mixing

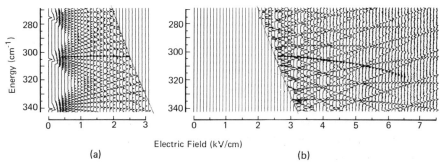

Fig. 5.17 Stark split Rydberg levels of sodium as detected by field ionization. Stepwise two-photon absorption populates the Stark split levels. (a) When the Stark field is strong enough to ionize all of the atoms before the detector gate opens, the resonances disappear. (b) When the detector gate opens when the laser pulse is applied and *closes* when it formerly opened, the resonances reappear when the ionization rate becomes fast enough to produce detectable ion signals in the limited detection time. At the highest fields, field ionization broadens the resonances (from Ref. [7]).

the triplet series of the alkaline earths can be identified and studied. By choosing different long-lived states as the initial level, one can determine the fractional parentage of high-lying levels. By gating the ionization detection electronics at different times, one can measure lifetimes and decay paths. The multiply resonant nature of the stepwise interaction and the use of coherent excitation processes for populating the initial level results in a large and easily identified ionizational signal. Stepwise processes of this sort are being developed for the separation of uranium isotopes on a commercial scale. Multichannel quantum defect theory can explain many features of the high-lying states of complex atoms which this technique opens to study [32]. The energy crisis fuels a burning interest in this area of laser spectroscopy.

5.4 NONLINEAR MIXING

Associated with each of these multiquantum absorption processes is a nonlinear polarization capable of generating coherent radiation. Measuring the intensity of that radiation as a function of one or more laser frequencies can provide more information about the highly excited energy levels than do the ionization or fluorescence experiments. These resonant nonlinear mixing processes can also be used to generate radiation useful for linear spectroscopy, and this topic will be discussed in Chapter 7.

The radiated signal need not be at the frequency of the transition; in fact, even quantum transitions cannot be at such frequencies. For one-, two-, and three-photon absorptions, the nonlinear polarization is correctly described by the third-order susceptibility in Eq. (2.9.8). The argument of Chapter 4 implies that nonlinear polarizations exist at the input frequencies and at threefold sums and differences of the inputs. Some of the four-wave mixing processes that can access multiquantum resonances when there are two different input frequencies are shown in Fig. 5.18. The multiple resonance conditions lead as usual to the largest signals.

Two-photon absorption resonances have been studied extensively by this technique. In analogy to the treatment of the Raman resonance in Chapter 4, one can write the coupling Hamiltonian and dipole operator in terms of a two-photon tensor [33],

$$\alpha_{\alpha\beta}^T(\omega_j, \omega_k) = \frac{\langle g|\tilde{\mu}_\alpha|n\rangle\langle n|\tilde{\mu}_\beta|t\rangle}{\Omega_{ng} + \omega_j} + \frac{\langle g|\tilde{\mu}_\beta|n\rangle\langle n|\tilde{\mu}_\alpha|t\rangle}{\Omega_{ng} + \omega_k} \quad (5.4.1)$$

in which case the two-photon resonant third-order nonlinear susceptibility tensor has the form

$$\chi_{\alpha\beta\gamma\delta}^T(-\omega_3, \omega_0, \omega_1, \omega_2) = -\frac{\mathcal{N}}{12\hbar}\sum_t \left\{ \frac{\alpha_{\beta\delta}^T(\omega_0, -\omega_1)\alpha_{\alpha\delta}^T(-\omega_3, \omega_2)}{\Omega_{tg} - (\omega_0 + \omega_1) - iT_2^{-1}} \right.$$
$$\left. + \frac{\alpha_{\beta\delta}^T(\omega_0, \omega_2)\alpha_{\alpha\delta}^T(-\omega_3, -\omega_1)}{\Omega_{tg} - (\omega_0 + \omega_2) - iT_2^{-1}} + \frac{\alpha_{\gamma\delta}^T(\omega_1, \omega_2)\alpha_{\alpha\beta}^T(-\omega_3, -\omega_0)}{\Omega_{tg} - (\omega_1 + \omega_2) - iT_2^{-1}} \right\}. \quad (5.4.2)$$

Note that the possibility of resonances at the input and output frequencies are contained in $\alpha^T(\omega_j, \omega_k)$. If all the frequency arguments are positive, the

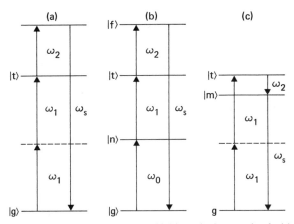

Fig. 5.18 Nonlinear mixing processes accessing highly excited energy levels. (a) A sum generation process resonant on the two-photon transition $|g\rangle \to |t\rangle$; (b, c) multiply resonant four-wave mixing processes. (b) The triple resonant process produces a large resonant nonlinearity, but linear absorption in the sample may reduce the signal. (c) The four-wave mixing process can spontaneously produce frequencies ω_2 and ω_s.

5.4 Nonlinear Mixing

output frequency is $\omega_3 = \omega_0 + \omega_1 + \omega_2$ and two-photon resonances occur for $\Omega_{tg} = \omega_0 + \omega_1$, $\Omega_{tg} = \omega_0 + \omega_2$, and $\Omega_{tg} = \omega_1 + \omega_2$. If one or more of the frequency arguments is negative, some of the two-photon resonances become Raman resonances, and the output polarization occurs at one of the frequencies discussed in Chapter 4. The Raman resonant and two-photon resonant nonlinear susceptibilities behave analogously in every detail. In materials where Raman and two-photon resonances can be observed simultaneously, four-wave mixing spectroscopy allows the two-photon absorption cross section to be normalized to the Raman cross section.

Angular momentum selection rules can be employed to separate the signal due to the desired resonance from other effects. In particular, there can be no output at $\omega_3 = 2\omega_1 + \omega_2$ when the waves at ω_1 and ω_2 are circularly polarized in the same sense. By making the wave at ω_1 right circularly polarized and the wave at ω_2 left circularly polarized, Wynne et al. obtained a tunable nonlinear mixing signal at $2\omega_1 + \omega_2$ resonant with an S → D transition at $\Omega_{tg} = 2\omega_1$ without spurious signals at $3\omega_1$, $3\omega_2$, and $2\omega_2 + \omega_1$. Liao and Bjorklund used the angular momentum selection rules near S → S and S → D two-photon transitions to produce a rotation in the plane of polarization of a probe wave due to a circularly polarized pump beam [34]. Their spectroscopic technique is the two-photon analog of Raman-induced Kerr effect spectroscopy, with the resonance occurring at $\Omega_{tg} = \omega_1 + \omega_2$ rather than at $\Omega_{rg} = \omega_1 - \omega_2$.

Ordinarily, absorption at the input and output frequencies partially obscures the increase in generated intensity due to an input or output frequency resonance in the nonlinear susceptibility. When a highly excited state couples more strongly to a nearby level than to the ground state, tuning that nearby level into two-photon resonance can emphasize the output frequency resonance in four-wave mixing. An experiment to study autoionizing states of strontium by this method is shown in Fig. 5.19. The frequency $2\omega_1$ was tuned into resonance with a two-photon resonant state with configuration $5p^2$ $^1D_2$ or $5s5d$ $^1D_2$, and the ω_2 frequency was scanned through the 4d4f autoionizing resonance at 1867 Å and other VUV levels [13].

Autoionizing resonances are narrow perturbations in the density of continuum states that occur when the energy of a bound level for one configuration equals that of a continuum level for another configuration. The line shapes that result are more complex than the Lorentzian resonances of truly discrete levels; an extensive theory has been developed by Fano [35]. In the case of strontium, the first ionization limit at 45,932 cm^{-1} corresponds to the limit of the series 5snp $^1P_1^0$. The excited electron then breaks free of the atom, but remains in a p-symmetry wave function. The second optically active electron continues to be in its lowest (5s) orbital. The overall symmetry

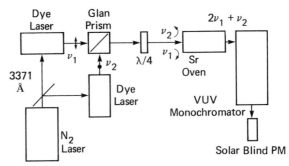

Fig. 5.19 A nonlinear mixing spectrometer. The dye laser beams are combined with orthogonal linear polarization at the glan prism and converted into opposite senses of circular polarization at the λ/4 plate. One dye laser is tuned to a convenient two-photon resonance, while the other is scanned across autoionizing resonances. The enhancement of the sum frequency amplitude created in the strontium oven is detected by a photomultiplier insensitive to the visible laser radiation (from Ref. [13]).

of the state is $^1P_1^0$. There exist nominally bound states of the same overall symmetry above this ionization limit. One such state is denoted 4d4f $^1P_1^0$ with energy 53,546 cm^1. Both optically active electrons are excited above their lowest energy orbitals into bound levels. The ionization limits for this configuration would leave a Sr$^+$ core in an excited state. The configuration interaction, however, mixes the 4d4f bound state with the 5sεp continuum state of the same energy allowing the "bound" state to decay by autoionization. A narrow perturbation in the density of continuum states results—the autoionizing resonance.

Two parameters are necessary to describe the lineshapes of the autoionization resonances observed in linear absorption. The strength of the configuration interaction V_E defines the linewidth, which can be quite narrow. Fano's q parameter is the ratio of the matrix elements coupling the ground state to the continuum and discrete portions of the final autoionizing state,

$$q_{fg} = \frac{\langle f_c|\tilde{\mu}|g\rangle}{\pi V_E^* \langle f_d|\tilde{\mu}|g\rangle}, \quad (5.4.3)$$

where $|f_c\rangle$ and $|f_d\rangle$ are the continuum and discrete wave functions coupled by the configuration interaction into the wave function for the autoionizing state. In terms of these parameters, the Fano–Beutler line-shape profile for linear absorption is

$$\chi''(\omega) \propto \frac{[q_{fg} - (\hbar\Omega_{fg} - \hbar\omega)/(\pi|V_E|^2)]^2}{1 + (\hbar\Omega_{fg} - \hbar\omega)^2/(\pi^2|V_E|^4)} \quad (5.4.4)$$

which is similar to the line shapes of CARS resonances.

5.4 Nonlinear Mixing

At exact two-photon resonance, the line shape describing four-wave mixing depends upon the output frequency $\omega_3 = 2\omega_1 - \omega_2$ and upon two Fano q parameters—one for the ground state and one for the two-photon resonant state $|t\rangle$,

$$|\chi^{(3)}|^2 \propto \frac{q_{fg}q_{ft} + [q_{fg} + q_{ft} - (\hbar\Omega_{fg} - \hbar\omega_3)/(\pi|V_E|^2)]^2}{1 + [(\hbar\Omega_{fg} - \hbar\omega_3)/(\pi|V_E|^2)]^2}. \quad (5.4.5)$$

Line shapes obtained experimentally by Armstrong and Wynne along with theoretical fits obtained by adjusting q_{ft}, the one free parameter in the theory, [13] are shown in Fig. 5.20. Similar measurements with ionization detection yield the line shape of (5.4.4) with $q = q_{ft}$ at exact two-photon resonance.

The sum generation intensity depends on the coherence length of the interaction as well as upon the nonlinear susceptibility. In the previously mentioned experimental studies, care was taken to ensure that the coherence length did not vary rapidly with frequency. The variation of the coherence length, however, contains information on the index of refraction of the medium and hence on the relative oscillator strengths of transitions between the input and output frequencies. For constant nonlinear susceptibility, the intensity produced by four-wave mixing with collinear plane wave inputs is

$$I \propto \sin^2(\Delta k l)/(\Delta k l)^2, \quad (5.4.6)$$

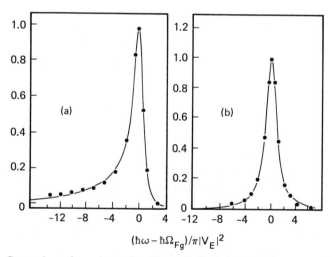

Fig. 5.20 Comparison of experimental and theoretical line shapes for four-wave mixing in the vacuum UV at an autoionization resonance. The atomic system was strontium; the solid line is the experiment while the dots represent the theory. (a) The 5p² resonance with $q_{fg} = -3.5$, $q_{ft} = -0.6$; (b) the 5s5d resonance with $q_{fg} = -3.5$, $q_{ft} = 2.1$ (from Ref. [13]).

where l is the sample length and

$$\Delta k = |\mathbf{k}_S - \mathbf{k}_0 - \mathbf{k}_1 - \mathbf{k}_2| = \frac{1}{c}\left|n(\omega_s)\omega_s - n(\omega_0)\omega_0 - n(\omega_1)\omega_1 - n(\omega_2)\omega_2\right|$$

$$\approx \frac{\mathcal{N}}{c}\left|\frac{f_{ng}\omega_s}{\Omega_{ng}^2 - \omega_s^2} - f_{pg}\left(\frac{\omega_0}{\Omega_{pg}^2 - \omega_0^2} + \frac{\omega_1}{\Omega_{pg}^2 - \omega_1^2} + \frac{\omega_2}{\Omega_{pg}^2 - \omega_2^2}\right)\right|. \tag{5.4.7}$$

Bjorklund has shown that it is generally possible to achieve $\Delta k = 0$ at one set of input frequencies when the output lies above the main resonance line [36]. The second equal sign in (5.4.7) applies when the output frequency $\omega_s = \omega_0 + \omega_1 + \omega_2$ is near a transition between ground state $|g\rangle$ and excited state $|n\rangle$ with oscillator strength f_{ng} and resonant frequency Ω_{ng}. The oscillator strength of the main resonant line is f_{pg} and the frequency is Ω_{pg}. Wynne has shown that the ratio of oscillator strengths f_{pg}/f_{ng} can be determined to a few percent by finding combinations of frequencies where perfect phase matching occurs. [37].

The importance of this method can be judged by the fact that previous methods of measuring oscillator strength ratios often have accuracies no better than $\pm 50\%$. This method determines the oscillator strength in terms of easily measured frequencies and is insensitive to atomic density and linewidths.

5.5 APPLICATIONS

A number of applications of multiquantum absorption phenomena are self-evident from the previous discussion. For example, the angular momentum selection rules can be used to identify highly excited electronic states of atoms pretty much as the selection rules for linear absorption and emission have been used for lower states. In the case of multiphoton ionization of molecules, the selection rules are less restrictive, but El-Sayed et al. have shown that the ratio of ion signals produced in different polarizations by two-photon resonant three-photon ionization still characterizes the molecular symmetry [38]. For two-photon resonant states with less symmetry than the molecule as a whole, the ratio of the circular to linear polarization cross sections must be $\frac{3}{2}$. Totally symmetric states show a ratio less than $\frac{3}{2}$, often much less. El-Sayed et al. used this information to assign the lowest-energy Rydberg states of hexatriene, pyridine, and pyrazine.

Other applications are surrounded by secrecy. Government agencies and major energy companies are developing separation schemes for uranium isotopes based on multiquantum ionization and multiquantum dissociation, but their spokesmen have been forbidden even to speak of uranium, much less the details of the process. It is clear that pilot plants presently under

5.5 Applications

construction will use one or more of these schemes to make enriched uranium for use in light water reactors from the tailings of more conventional enrichment processes. The laser isotope separation schemes are applicable to a wide range of elements including the other actinides and species used as tracers in biological experiments. Even more intriguing is the possibility of using isotope selective photodissociation or photoionization to concentrate the ^{14}C in a sample to improve the accuracy of radiocarbon dating.

Resonant multiphoton ionization can be used to detect single atoms of the resonant species in the presence of 10^{19} or so nonresonant atoms. Present-day lasers are capable of ionizing all of the atoms in the periodic table—except possibly He and Ne—with 95% efficiency [39]. A typical apparatus of this sort used to detect cesium atoms is shown in Fig. 5.21. The laser quanta have energy equal to the second resonance line, and this energy is also sufficient to ionize atoms reaching that level. Excited state ionization cross sections are typically 10^{-17} cm^2, requiring an energy density for saturation of only 100 mJ/cm^3. The liberated ions trigger an electrical signal in a modified Gieger tube or proportional counter. Such techniques can search for truly rare atoms, such as the Kr^{81} produced from Br^{81} by solar neutrinos.

Stepwise multiquantum absorption processes can produce populations of highly excited Rydberg atoms with unique properties. The field ionization threshold for such atoms varies rapidly with principal quantum number. It is possible to set the ionizing field to a value where it will not ionize the initially populated state at all, but will ionize the next higher level with 100% efficiency. Microwave or far-IR radiation resonant with the transition between these two levels is readily absorbed because of the enormous transition dipole moments of these loosely bound states. Once a Rydberg atom has

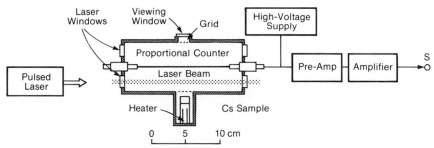

Fig. 5.21 Detecting single cesium atoms by resonant multiphoton ionization. The laser populates an excited state that is readily photoionized. The liberated charges trigger an electrical signal in the atmospheric pressure proportional counter. Present-day lasers can saturate one- and two-photon transitions between discrete levels as well as most one- or two-photon ionization processes from excited states. The detection efficiency for single atoms thus approaches 100%, (from Ref. [39]).

made such an upward transition, it can be ionized. Thus, it is now possible to construct resonant detectors for IR and microwave radiation with near unity quantum efficiency. Stark and Zeeman shift effects allow some tunability.

The Rydberg states are so delicate and have such large transition moments that the blackbody radiation in an ambient temperature apparatus can detectably shorten their lifetimes and shift the energies. Ultimately it may be possible to use these effects to define a quantum-electronic temperature scale referenced to well-defined fundamental constants [40].

REFERENCES

1. G. Grynberg, B. Cagnac, and F. Biraben, Multiphoton resonant processes in atoms, *in* "Coherent Nonlinear Optics" (M. S. Feld and V. S. Letokhov, eds.) (Topics in Current Physics 21), pp. 111–164. Springer-Verlag, Berlin, 1980.
2. N. Bloembergen and M. D. Levenson, Doppler-free two-photon absorption spectroscopy, *in* "High Resolution Laser Spectroscopy" (K. Shimoda, ed.) (Topics in Applied Physics 13), pp. 315–369. Springer-Verlag, Berlin, 1976.
3. M. M. Salour, Ultrahigh-resolution two-photon spectroscopy in atomic and molecular vapors, *Ann. Phys. (NY)* **111**, 364–503 (1978).
4. J. H. Eberly and P. Lambropoulos, eds., "Multiphoton Proccess," Wiley, New York, 1978.
5. P. Lambropoulos, Topics on multiphoton processes in atoms, *Advan. At. Mol. Phys.* **12**, 87–163 (1976).
6. J. S. Bakos, *Advan. Electron. Electron Phys.* **36**, 57 (1974).
7. M. L. Zimmerman, M. G. Littman, M. M. Kash, and D. Kleppner, *Phys. Rev. A* **20**, 2251–2275 (1979).
8. C. D. Cantrell, ed., "Multiphoton Excitation and Dissociation of Polyatomic Molecules." Springer-Verlag, New York (to be published).
9. J. H. Eberly and B. Karczewski, eds., "Multiphoton Bibliography 1970–1976." Univ. of Rochester, 1977. (Supplements Issued 1979, 1980 by Univ. of Rochester and Univ. of Colo NBS LP-92.)
10. C. D. Cantrell, V. S. Letokhov, and A. A. Makarov, Coherent excitation of multilevel systems by laser light, *in* "Coherent Nonlinear Optics" (M. S. Feld and V. S. Letokhov, eds.), pp. 165–270. Springer-Verlag, Berlin, 1981.
11. J. R. Ackerhalt and J. H. Eberly, Coherence versus incoherence in stepwise laser excitation on atoms, *Phys. Rev. A* **14**, 1705–1710 (1976).
12. P. A. Schulz, A. S. Sudbo, D. J. Krajnovich, H. S. Kwak, Y. R. Shen, and Y. T. Lee, *Ann. Rev. Phys. Chem.* **30**, 379 (1979); V. S. Letokhov and C. B. Moore, *Sov. J. Quant. Electron.* **6**, 254 (1976).
13. J. A. Armstrong and J. J. Wynne, The nonlinear optics of autoionizing resonances, in "Nonlinear Spectroscopy" (N. Bloembergen ed.), pp. 152–169. North-Holland Publ., Amsterdam, 1977; R. T. Hodgson, P. O. Sorokin, and J. J. Wynne, *Phys. Rev. Lett.* **32**, 343 (1974).
14. J. H. Eberly and S. V. O'Neil, Coherence versus incoherence: Time independent rates for resonant multiphoton ionization, *Phys. Rev. A* **19**, 1161–1168 (1979).

References

15. P. Lambropoulos, C. Kikuch, and R. K. Osborn, *Phys. Rev.* **144**, 1081 (1966); S. N. Dixit, P. Zoller, and P. Lambropoulos, Non-Lorentzian laser line shapes and the reversed peak asymmetry in double optical resonance, *Phys. Rev. A* **21**, 1289–1296 (1980), and references therein.
16. I. I. Sobel'man, "An Introduction to the Theory or Atomic Spectra." Pergamon, Oxford, 1972.
17. T. W. Hänsch, Nonlinear high resolution spectroscopy of atoms and molecules, *in* "Nonlinear Optics" (N. Bloembergen, ed.) (Proc. Int. School, Enrico Fermi Course 64), pp. 17–86. North-Holland Publ., Amsterdam, 1977.
18. W. K. Bischell, P. J. Kelly, and C. K. Rhodes, *Phys. Rev. Lett.* **34**, 300 (1975); *Phys. Rev. A* **13**, 1817, 1829 (1976).
19. C. Wieman and T. W. Hänsch, *in* "Laser Spectroscopy III" (J. L. Hall and J. L. Carlsten, eds.) Springer Series in Optical Sciences 7), p. 39. Springer-Verlag, Berlin, (1977).
20. S. A. Lee, J. Helmcke, J. L. Hall, and B. P. Stoicheff, Doppler-free two-photon transitions in Rydberg levels: Convenient, useful, and precise reference wavelengths for dye lasers, *Opt. Lett.* **3**, 141–143 (1978).
21. B. P. Stoicheff and E. Weinberger, Frequency shifts, line broadenings and Phase-Interference effects in Rb**–Rb collisions, measured in Doppler-free two-photon spectroscopy, *Phys. Rev. Lett.* **44**, 733 (1980).
22. S. E. Wheatley, P. Agostini, S. N. Dixit, and M. D. Levenson, Saturation effects in resonant three-photon ionization of potassium, *Physica Scripta* **18**, 177–181 (1978), and references therein.
23. P. Agostini, C. Barjot, J. F. Bonnal, G. Mainfray, C. Manus, and J. Morellec, *IEEE J. Quant. Electron.* **QE-6**, 783 (1970).
24. M. R. Cervenan and N. R. Isenor, *Opt. Commun.* **10**, 280 (1974); **13**, 175 (1975).
25. P. Agostini, A. J. Georges, S. E. Wheatley, P. Lambropoulos, and M. D. Levenson, Saturation effects in resonant three-photon ionization of sodium with nonmonochromatic field, *J. Phys. B* **11**, 1733–1747 (1978).
26. J. H. Eberly, Extended two-level theory of exponential index of multiphoton processes, *Phys. Rev. Lett.* **42**, 1049–1052 (1979).
27. G. Petite, J. Morellec, and D. Normand, Resonant multiphoton ionization of caesium atoms, *J. Phys.* **40**, 115–128 (1979).
28. D. E. Nitz, A. V. Smith, M. D. Levenson, and S. J. Smith, Bandwidth induced reversal of asymmetry in optical double resonant amplitudes, *Phys. Rev. A* **24**, 288–293 (1981).
29. S. H. Autler and C. H. Townes, *Phys. Rev.* **100**, 703 (1955).
30. J. A. Armstrong, J. J. Wynne, and P. Escherick, Bound, odd parity $J = 1$ spectra of the alkaline earths: Ca, Sr, and Ba, *J. Opt. Soc. Amer.* **64**, 211–230 (1979).
31. J. A. Armstrong, P. Escherick, and J. J. Wynne, Bound even-parity $J = 0$ and 2 spectra of Ca: A multichannel quantum defect analysis, *Phys. Rev. A* **15**, 180–196 (1977).
32. U. Fano, *J. Opt. Soc. Amer.* **65**, 979 (1975), and references therein.
33. P. D. Maker and R. W. Terhune, *Phys. Rev.* **137**, A801 (1965).
34. P. F. Liao and G. C. Bjorklund, Polarization rotation effects in atomic sodium vapor, *Phys. Rev. A* **15**, 2009–2018 (1977).
35. U. Fano, *Phys. Rev.* **125**, 1866 (1961).
36. G. C. Bjorklund, J. E. Bjorkholm, P. F. Liao, and R. H. Storz, Phase matching of two-photon resonant four-wave mixing processes in alkali metal vapors, *Appl. Phys. Lett.* **39**, 729–732 (1976).
37. J. J. Wynne and R. Beigang, Accurate relative oscillator strength determination by phase matching, *J. Opt. Soc. Amer.* **70**, 625 (1980).

38. D. H. Parker, J. O. Berg, and M. A. El-Sayed, Multiphoton ionization spectroscopy of polyatomic molecules, *in* "Advances in Laser Chemistry" (A. H. Zewail, ed.) (Springer Series in Chemical Physics 3) pp. 320–335. Springer-Verlag, Berlin, 1978.
39. M. H. Nayfeh, Laser detection of single atoms, *Amer. Scientist* **67**, 204–213 (1979), and references therein; G. S. Hurst, M. G. Payne, S. D. Kramer, and J. P. Young, *Rev. Mod. Phys.* **51**, 767 (1979).
40. S. Haroche, C. Fabre, P. Goy, M. Gross, and J. M. Raimond, Rydberg states and microwaves: High resolution spectroscopy, masers and superradiance, *in* "Laser Spectroscopy IV" (H. Walther and K. W. Rothe, eds.) (Springer Series in Optical Science **21**), Springer-Verlag, Berlin, 1979.

Chapter 6
OPTICAL COHERENT TRANSIENTS

Perhaps the most elegant of the nonlinear laser spectroscopy techniques are those which exploit the similarity of all two-level quantum systems to produce optical effects analogous to the pulsed NMR phenomena observed at radio frequencies [1]. Central to these techniques is the ability to prepare ensembles of quantum systems coherently and interrogate them in a time comparable to the transverse relaxation. Present-day laser techniques can span the time domain from 10 ms to 1 ps, but faster and slower processes still pose difficulties. In this section we shall refer extensively to the formalism developed in Sections 2.4–2.6, emphasizing the radiating polarizations driven by the ensemble. Different techniques detect the radiated field in different ways as outlined in Section 2.9. By emphasizing the effective two-level model, this treatment neglects many interesting and potentially useful phenomena. For these, the reader is referred to more specialized publications [2–7].

6.1 THE OPTICAL FREE-INDUCTION DECAY

Imagine that an external field with Rabi frequency χ_0 began exciting a homogeneously broadened ensemble of two-level system at $t = -\infty$, so that by $t = 0$ the Bloch–Feynman vector has reached the steady-state values in Eq. (2.4.21) and Fig. 2.5. If the driving field is then suddenly extinguished, the steady-state values of u, v, and w can be inserted in Eq. (2.4.20) to find the components of the **R** vector at later times. According to (2.4.20), the ensemble does not immediately return to thermal equilibrium, but rather continues to radiate a wave according to (2.5.3). This decaying optical field radiated coherently by an undriven system is the simplest optical free-induction decay (FID) [1, 8]. It is completely analogous to the singing of a violin string that continues after the bow has been removed.

Another view of the FID is that the ensemble has been absorbing radiation since $t = -\infty$. That absorption could be described by separating the amplitude transmitted through the sample into two terms: the incident amplitude and the amplitude of the wave radiated by the sample which interferes destructively with the incident amplitude and reduces the intensity. This reduction is the essence of absorption. When the incident driving field is suddenly turned off, the field radiated by the sample continues for a time $\sim T_2$ and is detected as the optical free induction decay. If the absorption is not saturated, the magnitude of this FID amplitude scales linearly with the Rabi frequency of the incident radiation. Thus, this amplitude is technically termed the "linear" or "first-order" FID. Note that nothing in our description so far requires that the Rabi frequency be linear in the incident amplitude; first-order FID effects have been seen with nonlinear coupling to the ensemble. [9–11].

If the ensemble had been excited by a short pulse of radiation between times $t = -t_p$ and $t = 0$ with $t_p \ll T_2 \le T_b$, Eq. (2.4.17) would have given u_0, v_0, and w_0. A similar first-order FID would have resulted which would be analogous to the ringing of a bell that had been struck by a hammer. For such impulsive excitation, linear systems theory implies that the time evolution of polarization reflects the Fourier transform of the linear susceptibility,

$$P^{(1)}(t) \propto \int_{-\infty}^{\infty} \chi^{(1)}(\omega) e^{-i\omega t} \, d\omega, \tag{6.1.1}$$

so long as the pulse width is much shorter than the inverse of the width of the absoprtion spectrum. Thus, beats can be heard between the sounds of two bells struck with the same hammer, and a bell made of lead makes a sound that decays rapidly because of the large bandwidth of its resonance. The frequency of the first-order FID is that of the transition—not the driving field, just as the note of a bell depends on the characteristics of the bell and not the clapper.

The situation for an inhomogeneously broadened system is a little more complicated. If the driving Rabi frequency is sufficient to saturate the transition even a little, the line-shape function becomes intensity dependent and a new component appears in the FID signal. This new component is termed the "third-order" FID because its amplitude is proportional to the product of the depth of the hole (which varies as χ^2) and χ [2]. The third-order FID continues for a time proportional to T_2, even though the first

6.1 The Optical Free-Induction Decay

order FID decays in a time T_2^* equal roughly to the inverse of the inhomogeneous linewidth. Since $T_2 \gg T_2^*$ for most cases, one can distinguish the first- and third-order effects and both experimentally and theoretically.

The third-order FID polarization for steady-state excitation ending at $t = 0$ can be calculated by subtracting the terms linear in the Rabi frequency, χ_0, from (2.4.21) and using the result as the coefficients u_0, v_0, and w_0 in (2.4.20). The total polarization is then obtained by integrating (2.5.3) over

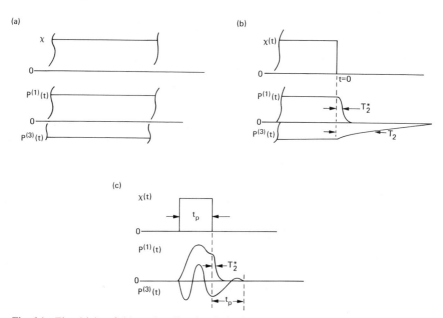

Fig. 6.1 The driving fields and radiated polarizations for steady-state absorption and free induction decays (FIDs) with steady-state and pulsed preparation. (a) In the steady-state case, the driving fields are constant in time for $-\infty < t < \infty$ and thus the Rabi frequency χ is also constant in time. The resonance of the sample linearly proportional to χ is a polarization density $P^{(1)}$ that radiates fields that interfere destructively with the incident fields, thus reducing the transmitted power. If the absorption is slightly saturated, there is also a dielectric polarization density $P^{(3)}$ proportional to χ^3 which radiates a field that interferes constructively with the transmitted driving fields thus apparently reducing the absorption. (b) If the driving fields are turned off suddenly at $t = 0$, the sample continues to radiate until the ensemble dephases. The polarization $P^{(1)}$ due to the linear response decays to roughly $1/e^2$ of its initial value in the inhomogeneous dephasing time T_2^* which is inversely proportional to the inhomogeneous linewidth. The wave radiated in this way for $t > 0$ is the first-order FID. The third-order polarization density $P^{(3)}$ decays more slowly reaching $1/e$ its initial value in a time $T_2/2$ related to the inverse of the homogeneous linewidth. The resulting signal is the third-order FID which is obviously shifted in phase from the first-order FID amplitude. (c) Pulsed excitation causes optical nutation for $-t_p < t < 0$ followed by FIDs as before. The first-order signal again decays rapidly to zero while the third-order amplitude can show an oscillatory time dependence. A theorem by Schenzle et al. [13] shows that the third-order FID decays to zero in time t_p.

the inhomogeneous distribution of transition frequencies,

$$P^{(3)}(t) = \frac{-\{-1\}G(\omega-\Omega_{ab})\mu_{ab}\chi_0^3 T_b e^{-i\omega t}}{T_2}$$

$$\times \int_{-\infty}^{\infty} \frac{\mathcal{N} e^{(i\Delta t - t/T_2)}}{(\Delta + iT_2^{-1})(\Delta^2 + T_2^{-2} + \chi_0^2 T_b/T_2)} d\Delta$$

$$= \frac{\pi i\{-1\}\mathcal{N} G(\omega-\Omega_{ab})\mu_{ab}\chi_0^3 T_b \exp(-(t/T_2)[1+\sqrt{\chi_0^2 T_b/T_2 + 1}])}{T_2^{-1} + \chi_0^2 T_b + (T_2^{-2} + \chi_0^2 T_b/T_2)^{1/2}} e^{-i\omega t}$$

(6.1.2)

where $G(\omega - \Omega_{ab})$—the distribution function—is assumed to be so slowly varying that it may be taken out of the integral. One should note that the frequency ω appearing in (6.1.2) is the frequency of the field used to prepare the ensemble during $-\infty < t < 0$. In frequency-switching experiments, the field radiated by this polarization interferes at the detector with a frequency-shifted wave from the same laser and gives rise to a heterodyne beat signal [2]. The time evolution of these amplitudes is illustrated in Fig. 6.1. The

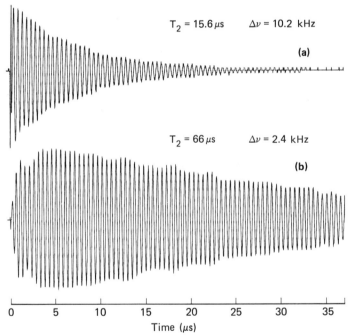

Fig. 6.2 Optical heterodyne detected third-order free-induction decay signals in $Pr^{+3}:LaF_3$. The electrical signal oscillates because the local oscillator provided to enhance the sensitivity is shifted in frequency from the third-order FID amplitude. The decay of the envelope yields the transverse relaxation time, which in this system increases as in the lower trace, when appropriate rf irradiation is applied. The homogeneous linewidth indicated as $\Delta \nu$ is more than five orders of magnitude narrower than the inhomogeneously broadened absorption profile (from Ref. [12]).

decay rate of the FID amplitude in the weakly saturated regime $\chi_0^2 T_b \ll T_2^{-1}$ is $2T_2^{-1}$. Half of this decay rate represents inhomogeneous broadening due to the width of the hole burned during the preparation stage, and half represents the intrinsic transverse decay rate of the ensemble. The observation of FID decays requiring 70 μs or so in $Pr^{+3}:LaF_3$ (see Fig. 6.2) has been interpreted as implying a homogeneous linewidth of 2.4 kHz. Measurement of the FID decay rate as a function of temperature, pressure, etc. is one of the most sensitive means of studying dephasing processes [12].

6.2 OPTICAL NUTATION

When a driving field is suddenly applied at $t = 0$ to an ensemble at thermal equilibrium, the Bloch–Feynman vector does not immediately reach its steady-state value, but rather precesses around the pseudofield vector β until relaxation processes have a chance to act. The precession frequency is $\beta = (\Delta^2 + \chi^2)^{1/2}$, and Eq. (2.4.17) describes the behavior of the **R** vector when relaxation can be totally ignored. The polarization resulting from the precession of **R** is called an optical nutation signal. Inhomogeneous broadening tends to damp out the optical nutation, converting its time dependence from a sinusoid to a Bessel function: $J_0(\chi t)$. The sudden rise of $J_0(\chi t)$ at $t = 0$ results from the approximation that the inhomogeneous linewidth is infinite and cannot be observed experimentally [2].

Optical nutation signals are always observed superimposed on the transmitted laser beam. When that beam is Gaussian, and the diameter of the detector is larger than the beam, the electrical signal is obtained by averaging terms proportional to

$$\int_0^\infty \chi(r) J_0(\chi(r)t) r\, dr$$

over the Gaussian spatial dependence of $\chi(r)$. The effect of this average is to convert the $J_0(\chi t)$ dependence into $J_1(\chi t)$ dependence where χ should then be interpreted as the maximum Rabi frequency occurring in the center of the laser beam. The nonphysical discontinuity at $t = 0$ is eliminated. Measuring the Rabi frequency by finding the zero crossings of the nutation signal is the most accurate means of gauging transition dipole moments [2,4].

Relaxation adds tremendous complexity to the theory of optical nutation, and it is generally necessary to solve the equations of motion (2.4.19) numerically. Qualitative understanding can be obtained from an analytic approximation appropriate when $T_2 = T_1 = T_b = T$ and $\chi T \gg 1$. The leading terms for the polarization as averaged over the inhomogeneously broadened profile are

$$P(t) = \pi i \mathcal{N} G(\omega - \Omega_{ab}) \mu_{ab} \chi \left\{ w(0) e^{-t/T} J_0(\chi t) + \frac{2w^e}{T(\chi^2 + T^{-2})^{1/2}} \right\} e^{-i\omega t}, \quad (6.2.1)$$

where we have also assumed that $v(t = 0-) = u(t = 0-) = 0$ [2]. The first term is identical to the undamped nutation except that there is now an additional exponential damping factor. The second term reflects the onset of steady-state behavior. In the two-level atom approximation, w^e should be replaced by -1. The frequency ω appearing in (6.2.1) is the frequency of the field driving the nutation for $t > 0$.

A more complex form of optical nutation used to measure the lifetime of the excited state is illustrated in Fig. 6.3 [4]. An initial preparatory pulse—ideally with area π—disturbs the ensemble from thermal equilibrium. The magnitude of the nutation due to a subsequent pulse applied at time t is proportional to the population difference $w(t) = \rho_{bb}(t) - \rho_{aa}(t)$. For $t \to \infty$, $w(t)$ approaches its thermal equilibrium value. For shorter times, the nutation will be smaller, and a plot of the magnitude of the second nutation versus time will reveal the lifetime T_b. In such an experiment, one must take care to eliminate other transient effects and to minimize the effects of nonzero values of u and v left by the first pulse upon the second nutation signal. A result due to Schenzle et al. shows that the transverse components of **R** due to the first pulse will vanish identically in a time equal to the pulse length, if the system is inhomogeneously broadened [13]. The difference in maximum pulse heights of the two optical heterodyne-detected nutation signals indicated in Fig. 6.3 gives the time dependence of the population difference,

$$S(\infty) - S(t_d) \propto e^{-t_d/T_b}. \tag{6.2.2}$$

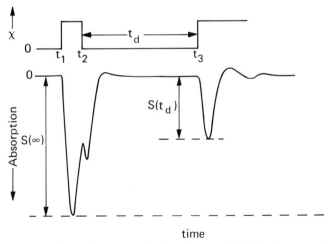

Fig. 6.3 The driving field pulses and optical nutation amplitudes in the two-pulse nutation technique. The variation of the height of the second nutation pulse with decay time t_d reveals the longitudinal relaxation rate.

6.3 THE PHOTON ECHO

When an ensemble is excited by two pulses of radiation separated by a time less than $\sim 3T_2$, the ensemble emits a third or echo pulse delayed from the second incident pulse by a time period equal to the time between the exciting pulses [1–6]. In inhomogeneously broadened systems, the photon echo can be a dramatic effect, appearing long after the ensemble coherence has apparently died away [14]. The photon echo can be visualized most easily in the strong field regime where the Rabi frequency is much larger than the inhomogeneous linewidth ($\chi T_2^* \gg 1$) and thus $\chi \gg \Delta$ for all members of the ensemble. In the ideal case, the area of the first pulse is $\Theta_1 = \frac{1}{2}\pi$ and that of the second pulse is $\Theta_2 = \pi$. We shall assume that the pulse lengths are much less than the relaxation times T_b and T_2 [4, 15]. We shall also employ the matrix transformations of Eqs. 2.4.16 and 2.4.17 to describe the effects of successive pulses on the **R** vectors. Note that for very short pulse lengths $\Theta = \beta t_p$.

The photon echo can be explained in terms of the vector model using the construction of Fig. 6.4. The first pulse rotates the **R** vectors for all the members of the ensemble to an orientation along the $\hat{2}$ axis at time $t = 0$,

$$\mathbf{R}_\Delta(0) = \{-1\}\hat{2} = \begin{bmatrix} 0 & 0 & 0 \\ 0 & 0 & 1 \\ 0 & -1 & 0 \end{bmatrix} \begin{pmatrix} 0 \\ 0 \\ -1 \end{pmatrix}. \quad (6.3.1)$$

The subscript Δ in (6.3.1) specifies the detuning of part of the ensemble from the incident frequency. At $t = 0$, the $\mathbf{R}_\Delta$ vectors for all the members of the ensemble are parallel, summing to a nonzero total $\mathbf{R}_T(0)$ which produces a first-order FID. As time goes on, however, the $\mathbf{R}_\Delta(t)$ vectors of the individual members of the ensemble precess around the $\hat{3}$ axis at their own rates according to (2.4.20). After a time T_2^*, the $\mathbf{R}_\Delta(t)$ vectors will be uniformly distributed around the 1-2 plane. Thus, the vector sum $\mathbf{R}_T$—which corresponds to the radiating polarization density—is zero. This situation is illustrated in Fig. 6.4b.

In the time interval between pulses, the projections of the $\mathbf{R}_\Delta(t)$ vectors on the 1-2 plane rotate and shrink in magnitude,

$$\mathbf{R}_\Delta(t) \sim \{-1\}\{\sin \Phi_\Delta(t)\hat{2} - \cos \Phi_\Delta(t)\hat{1}\}e^{-t/T_2} + w_\Delta \hat{3}. \quad (6.3.2)$$

At the top of Fig. 6.4 is a plot of $\Phi_\Delta(t) = \Delta t + \frac{1}{2}\pi$; the phase angle of $\mathbf{R}_\Delta(t)$ measured from the $\hat{1}$ axis for typical values of Δ. It is clear that while the ensemble described by $\mathbf{R}_T$ has dephased, subgroups with equal detuning continue to evolve coherently.

After time $t_d = t_2 - \frac{1}{2}t_p$, a second strong pulse is applied. The pulse length is t_p and the pulse area is ideally π. This second pulse has the effect of rotating the $\mathbf{R}(t)$ vectors *through an angle of 180° around the $\hat{1}$ axis*. This operation

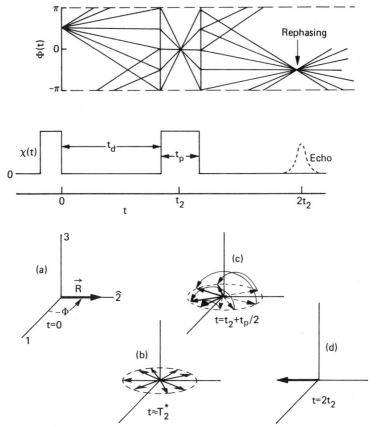

Fig. 6.4 The photon echo. (a) The first excitation pulse which ends at $t = 0$ excites the entire inhomogeneously broadened ensemble. A $\pi/2$ pulse would create a Bloch vector oriented as shown at $t = 0$. (b) Afterward, the phases of typical members of the ensemble increase or decrease at a rate characterized by the detuning Δ so that after an interval T_2^* the Bloch vectors are uniformly distributed around the 1-2 plane. (c) The second pulse—ideally with area π—reverses the phase of each member of the ensemble. After that pulse, the phases increase or decrease at the same rate as before, but after an interval t_2 all members of the ensemble are once again in phase with one another. (d) The nonvanishing Bloch vector that results; the dielectric polarization density that results radiates the photon echo pulse. The evolution of the phases of typical members of the ensemble is illustrated at the top.

is illustrated in Fig. 6.4c; the transformation can be represented analytically using (2.4.17),

$$\mathbf{R}(t_d + \tfrac{1}{2}t_p) = \begin{bmatrix} 1 & 0 & 0 \\ 0 & -1 & 0 \\ 0 & 0 & -1 \end{bmatrix} \mathbf{R}(t_d - \tfrac{1}{2}t_p). \quad (6.3.3)$$

6.3 The Photon Echo

The phase angles $\Phi_A(t)$ shown at the top of Fig. 6.4 are reversed by the π pulse. For $t > t_2 + \tfrac{1}{2}t_p$, the incident field is again zero and the $\mathbf{R}_A(t)$ vectors evolve according to (2.4.19) and (2.4.20) with the phase angle $\Phi_A(t)$ increasing or decreasing as in the time interval $0 < t < t_2 + \tfrac{1}{2}t_p$. However, the initial phase at $t = t_2 + \tfrac{1}{2}t_p$ is just the *negative* of the phase at $t = t_2 - \tfrac{1}{2}t_p$. Algebraically, the phase angles for $t > t_2 + \tfrac{1}{2}t_p$ are

$$\Phi_A(t) = \Delta(t - t_2 - \tfrac{1}{2}t_p) - \Phi_A(t_d). \tag{6.3.4}$$

The $\mathbf{R}_A(t)$ vectors evolve as in (6.3.2) with the new phase angles given by (6.3.4),

$$\mathbf{R}_A(t) = -\{\pm 1\}\{\sin(\Delta[t - 2t_2] - \tfrac{1}{2}\pi)\hat{2} \\ - \cos(\Delta[t - 2t_2] - \tfrac{1}{2}\pi)\hat{1}\}e^{-t/T_2} + w_A\hat{3}, \tag{6.3.5}$$

where we have evaluated $\Phi_A(t)$ for $t > t_2 + \tfrac{1}{2}t_p$.

After a time interval equal to t_2, all the $\mathbf{R}_A(t)$ vectors are again parallel and $\mathbf{R}_T$ is again nonzero. The inhomogeneously broadened ensemble has rephased as a whole producing a macroscopic polarization density

$$\mathbf{P}(t) = \{-1\}\mathcal{N}\boldsymbol{\mu}_{ab}\int_{-\infty}^{\infty} G(\Omega_{ab} + \Delta)\{\sin(\Delta[t - 2t_2] - \tfrac{1}{2}\pi) \\ - i\cos(\Delta[t - 2t_2] - \tfrac{1}{2}\pi)\}\,d\Delta\, e^{-t/T_2}e^{-i\Omega_{ab}t} \tag{6.3.6}$$

that peaks at time $t = 2t_2$. This rephasing is illustrated at the top of Fig. 6.4 and in Fig. 6.4d. The distribution function for transition frequencies is given in (6.3.6) as $G(\Omega_{ab} + \Delta)$. The rephasing ensemble radiates a pulse of radiation that is detected as the photon echo. As t_2 is varied, the amplitude of this polarization appearing at $t = 2t_2$ varies as

$$|P(2t_2)| \sim e^{-2t_2/T_2}. \tag{6.3.7}$$

Since the driving fields vanish except during the pulses, the photon echo decay rate is $2T_2^{-1}$, independent of the incident intensity. The shape of the echo pulse resembles two first-order FID pulses back to back in the strong field $(\chi T_2^* \ll 1)$ limit. More complex pulse shapes arise in the general case [2, 4].

The photon echo need not be excited by collinear beams. In particular, if the wave vector for the first pulse makes an angle ϕ with the wave-vector direction for the second (π area) pulse, the reversal of the phase factor causes the echo to be radiated at an angle of $-\phi$ with respect to the propagation direction of the second pulse as is illustrated in Fig. 6.5a. It should also be clear from Fig. 6.4 that a second π pulse centered at time $t = 3t_2$ would rephase the dipoles again causing a second echo pulse at time $t = 4t_2$, etc.

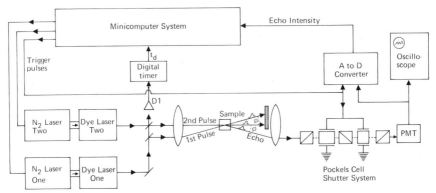

Fig. 6.5 Computerized photon echo apparatus with noncollinear excitation. The phase-matching condition resembles CARS in a condensed medium. The first and second pulses are incident upon the sample with an angle between them of ϕ; the echo is radiated at $-\phi$ as indicated. The entire system is controlled by a minicomputer which triggers the two nitrogen lasers at the times required for the pulses that excite the echo and opens the multistage Pockels-cell shutter system at the time the echo is expected. Jitter in the triggering of the nitrogen lasers requires an independent measure of the actual delay time which is obtained from detector D1. The effects of long period variations in the laser intensities are partially eliminated by normalizing the echo signal produced at one value of t_d to that produced at a standard time delay. In operation, alternate pulse pairs have the standard delay time, while the delay time for the remainder is scanned from zero until the echo disappears. Stimulated echoes and other effects can be produced either by splitting the output of one laser and storing a part of the pulse in an optical delay line or by adding a third independently triggered laser (after Ref. [18]).

Photon echoes occur under conditions far more general than those of the preceding derivation. The exciting pulses can have areas much less than $\Theta_1 = \frac{1}{2}\pi$ and $\Theta_2 = \pi$ and still produce an echo nearly as strong as in the ideal case. For arbitrary pulse areas and a Gaussian distribution of transition frequencies, $G(\Omega_{ab} + \Delta) = (\frac{1}{2}T_2^*/\sqrt{\pi})\exp(-\Delta^2 T_2^{*2})$, the photon echo polarization density in the strong field case is

$$\mathbf{P}(t) = \frac{\{-1\}\mu_{ab}}{2} \mathcal{N} \sin\Theta_1(1 - \cos\Theta_2)e^{-t/T_2}\exp\left(-\left[\frac{t - 2t_2}{2T_2^*}\right]^2\right)e^{-i\Omega_{ab}t}. \tag{6.3.8}$$

The Rabi frequency can be much smaller than the inhomogeneous broadening (i.e., $\chi T_2^* \ll 1$) and still there will be an echo. Only the magnitude of the echo and its pulse shape change when these conditions are relaxed, and the literature contains detailed calculations for the limiting cases [4]. At the lowest Rabi frequencies, the echo amplitude scales as χ^3, and the frequency of the echo pulse reflects the frequencies of the lasers used to excite it.

6.3 The Photon Echo

Whenever the transverse relaxation of the individual members of an inhomogeneously broadened ensemble is described by a single rate T_2^{-1}, the photon echo amplitude will decay at the same rate. Elastic collisions in gases can alter the velocity of a molecule without changing anything else. The effects of such collisions is to change the detuning rather than to add to the dephasing rate. Such collisions introduce a residual Doppler dephasing effective at $t = 2t_2$ [2, 16, 17]. When these effects are added to the theory of the photon echo, the decay of the echo amplitude for $t_2 \ll T_2$ scales as

$$|P(2t_2)| \sim \exp(-\Gamma \omega^2 t^3 \delta u^2 / 16c^2), \qquad (6.3.9)$$

where Γ is the rate of elastic collisions and δu is the mean change in axial velocity due to such collisions.

In three- and four-level systems where the frequency shift between levels in a group is less than the Rabi frequency, the photon echo amplitude can be modulated by effects analogous to quantum beats. It is not quite correct to treat the polarization radiated in such cases as the superposition of echoes due to the individual transitions; but the modulation of the peak echo amplitude has the $\cos(\Delta\Omega t_2)$ form predicted by such a model. The envelope function traced by the maximum of the photon echo as a function of t_2 in an experiment by Hartmann et al. on the $^3H_4 \rightarrow {}^3P_0$ transition in $Pr^{3+}:LaF_3$ is shown in Fig. 6.6. Note that the apparent rapid dephasing at short times is spurious, and is reversed at later times. By Fourier transforming this envelope pattern, Hartmann et al. obtained the spectrum in (6.6b) and were able to assign the splittings responsible for the complex beat pattern. In a similar experiment, Grossman et al. were able to extract dephasing rates for the individual transitions [19].

If a third pulse of area $\frac{1}{2}\pi$ is incident along the direction of the wave vector of the radiating polarization centered at $t = 2t_2$, the radiating polarization represented by the projection of $\mathbf{R}_T$ along the 2 axis can be converted into a population change. The third pulse acts to "store" the photon echo as a nonthermal population difference [20]. Since the population of the upper state can often be conveniently monitored by fluorescence or ionization, the change in population due to the stored photon echo can be detected. This technique is useful when the echo itself is too weak to detect or when light scattering by the sample cannot be suppressed. In practice, the experimental technique is to compare the fluorescence signal that results when the third pulse is incident at $t = 2t_2$ with that resulting when the third pulse occurs before or after the echo. If the difference between the two fluorescence signals varies as e^{-2t_2/T_2}, that difference is assigned to the stored echo. The concept of storing a polarization as a population difference is exploited more fully in the Ramsey fringe techniques of Section 6.5.

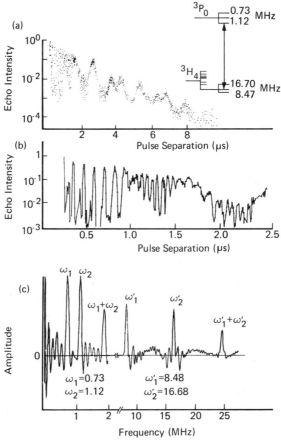

Fig. 6.6 Modulation of the photon echo intensity due to hyperfine level splittings in Pr:LaF$_3$. The energy levels involved are shown in the inset at the top, with the hyperfine splittings indicated in megahertz. (a) The experimentally observed echo intensity, with (b) the first 2.5 μs shown in an expanded scale. (c) Fourier transformation of the echo intensity envelope function results in the spectrum (from Ref. [18]).

6.4 THE STIMULATED ECHO

Several additional types of echo can be created when the ensemble is excited by three or more pulses of radiation [2–6]. The most adaptable of these new echoes is called the stimulated echo, and occurs at time $t = t_3 + t_2$ when the ensemble has been excited by $\frac{1}{2}\pi$ pulses at $t = 0$, $t = t_2$, and $t = t_3$ [17]. The evolution of the Bloch–Feynman vectors for this pulse sequence

6.4 The Stimulated Echo

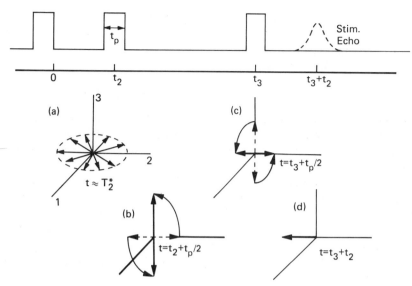

Fig. 6.7 The stimulated echo process. (a) Again the first $\pi/2$ pulse ends at $t = 0$ and the ensemble dephases in a time $\sim T_2^*$. The second pulse has area $\pi/2$ and "stores" the polarization pattern of the individual members of the ensemble as a detuning dependent population difference. (b) Geometrically, this process can be represented as a rotation of the Bloch vectors by $\pi/2$ around the $\hat{1}$ axis. The nonthermal population difference decays at the longitudinal relaxation rate, while the components of the Bloch vector remaining in the 1-2 plane decay at the (generally faster) transverse relaxation rate. (c) The third pulse rotates the Bloch vectors from the $\hat{3}$ axis into the 1-2 plane, and (d) after a time t_2 the ensemble rephases forming an echo as in the photon echo case.

is illustrated in Fig. 6.7. The first pulse creates a transverse component of $\mathbf{R}_T$ which dephases as in the photon echo. The second pulse rotates these dephased $\mathbf{R}_\Delta$ vectors into a plane containing the $\hat{3}$ axis; that is, it "stores" the coherence of the subensembles with detuning Δ as a pattern of population difference (see Fig. 6.7b). This nonthermal population difference decays toward equilibrium at the longitudinal rate T_b^{-1}, which can be much slower than the transverse decay rate. The third $\frac{1}{2}\pi$ pulse rotates the w_Δ pattern back into the $\hat{1}$-$\hat{2}$ plane (as in Fig. 6.7c), where the $\mathbf{R}_\Delta(t)$ vectors converge on a nonzero sum after a time interval equal to the interval between excitation pulses. The polarization corresponding to the rephased sum radiates the stimulated echo.

The second and third pulses of the stimulated echo sequence are analogous to a very long second pulse in the photon echo sequence, but the interesting physics of the stimulated echo takes place between these two pulses. In that interval, the transverse components of $\mathbf{R}_\Delta(t)$ can all relax to zero, yet there will be an echo. Only the longitudinal component matters, and that component can be very long-lived. Stimulated echos have been detected for $t_3 = 45$ min. In our simple two-level atom picture, the time dependence of the peak stimulated echo polarization is

$$|P(t_2 + t_3)| \propto \exp(-(t_3 - t_2)/T_b)\exp(-2t_2/T_2). \tag{6.4.1}$$

Thus, both the longitudinal and transverse relaxation can be measured with a single echo technique. The radiated frequency equals the transition frequency in the strong-field limit and the exciting laser frequency in the weak-field limit.

In order to make the subtle features of the stimulated echo more transparent, we shall derive the polarization using a simplified model in which the width of the exciting pulses are set to zero ($t_p \to 0$), but the pulse areas remain $\frac{1}{2}\pi$. This intrinsically implies $|\chi|T_2^* \gg 1$. To emphasize the spatial dependence of the polarization, we shall explicitly include the traveling-wave nature of the excitation pulses by writing the complex Rabi frequency defined in (2.4.2) as

$$\chi_\omega(t) = \sum_i (-\tfrac{1}{2}\pi \exp[i(\mathbf{k}_i \cdot \mathbf{r} + \phi_i)] e^{-i\omega t} \delta(t - t_i)); \tag{6.4.2}$$

the pseudofield vectors rotating at $\omega = \Omega_{ab}$ can be defined by (2.4.8) and (2.4.11) as

$$\boldsymbol{\beta}(t) = \tfrac{1}{2}\pi \sum_i (\cos(\mathbf{k}_i \cdot \mathbf{r} + \phi_i)\hat{1} + \sin(\mathbf{k}_i \cdot \mathbf{r} + \phi_i)\hat{2}) \delta(t - t_i) + \Delta\hat{3}. \tag{6.4.3}$$

With these definitions, the components of the $\mathbf{R}_\Delta$ vectors and the pseudofield vectors remain real.

If the ensemble begins in thermal equilibrium with $w = \{-1\}$, the effect of the first pulse at $t = t_1 = 0$ is to rotate the $\mathbf{R}$ vector into the $\hat{1}$-$\hat{2}$ plane with orientation perpendicular to $\boldsymbol{\beta}(t_1)$,

$$\mathbf{R}_\Delta(0+) = u\hat{1} + v\hat{2} = \{-1\}[-\sin(\mathbf{k}_1 \cdot \mathbf{r} + \phi_1)\hat{1} + \cos(\mathbf{k}_1 \cdot \mathbf{r} + \phi_1)\hat{2}]. \tag{6.4.4}$$

In the interval $0 < t < t_2$ the components of the $\mathbf{R}_\Delta$ vectors evolve according to (2.4.20),

$$\mathbf{R}_\Delta(t) = \{-1\}[-\sin(\mathbf{k}_1 \cdot \mathbf{r} + \phi_1)\cos\Delta t - \cos(\mathbf{k}_1 \cdot \mathbf{r}_1 + \phi_1)\sin\Delta t]e^{-t/T_2}\hat{1}$$
$$+ \{-1\}[\cos(\mathbf{k}_1 \cdot \mathbf{r} + \phi_1)\cos\Delta t - \sin(\mathbf{k}_1 \cdot \mathbf{r}_1 + \phi_1)\sin\Delta t]e^{-t/T_2}\hat{2} + w\hat{3}. \tag{6.4.5}$$

The second pulse rotates the $\mathbf{R}_\Delta(t_2-)$ vectors around the axis defined by the vector $\boldsymbol{\beta}(t_2)$ by an angle of $90°$. The transformation can be represented

6.4 The Stimulated Echo

algebraically by

$$\mathbf{R}_A(t_2+) = \begin{bmatrix} \cos^2(\mathbf{k}_2 \cdot \mathbf{r} + \phi_2) & \sin(\mathbf{k}_2 \cdot \mathbf{r} + \phi_2)\cos(\mathbf{k}_2 \cdot \mathbf{r} + \phi_2) & \sin(\mathbf{k}_2 \cdot \mathbf{r} + \phi_2) \\ \sin(\mathbf{k}_2 \cdot \mathbf{r} + \phi_2)\cos(\mathbf{k}_2 \cdot \mathbf{r} + \phi_2) & \sin^2(\mathbf{k}_2 \cdot \mathbf{r} + \phi_2) & \cos(\mathbf{k}_2 \cdot \mathbf{r} + \phi_2) \\ -\sin(\mathbf{k}_2 \cdot \mathbf{r} + \phi_2) & \cos(\mathbf{k}_2 \cdot \mathbf{r} + \phi_2) & 1 \end{bmatrix} \mathbf{R}_A(t_2-),$$

(6.4.6)

where t_2+ and t_2- refer to times just before and just after second pulse, and the Δ dependence vanishes because $\Delta t_p \ll 1$. The projection of $\mathbf{R}_A(t_2+)$ along the 3 axis is

$$w_A(t_2+) = \{-1\}\cos(\Delta t_2 - (\mathbf{k}_1 - \mathbf{k}_2) \cdot \mathbf{r} + \phi_1 - \phi_2)e^{-t_2/T_2}. \quad (6.4.7)$$

Since the transverse components of $\mathbf{R}_A$ present at this stage play no further role in the stimulated echo, they will henceforth be omitted. Note that Eq. (6.4.7) shows that the population difference has a sinusoidal dependence on the detuning from line center Δ. The coefficient of Δ encodes the interval between the first two pulses; this oscillatory dependence of w_A is the characteristic of the stimulated echo that allows the generation of greatly delayed echoes.

The spatial dependence of $w_A(t_2+)$ contains a Fourier component at the difference of the wave vectors of the exciting pulses. Those familiar with degenerate four-wave mixing or moving grating effects will recognize that such population gratings can scatter incident radiation. The "scattering" of the third or stimulating pulse will be responsible for the stimulated echo. Normally stimulated echo experiments are performed either with copropagating or counterpropagating beams for the two initial pulses. In the former case $\mathbf{k}_1 = \mathbf{k}_2 = \mathbf{k}$, the population difference is spatially uniform. An incident stimulating pulse will be "scattered" forward. For counterpropagating beams $\mathbf{k}_2 = -\mathbf{k}_1 = -\mathbf{k}$ and the grating with spacing $\pi/(2k)$ scatters the stimulated echo backward with respect to the direction of the third pulse.

In the time interval $t_2 < t < t_3$ the oscillatory component of $w_A(t)$ decays with the longitudinal decay rate

$$w_A(t) = w_A(t_2+)e^{-(t-t_2)/T_b} \quad (6.4.8)$$

and the third pulse at $t = t_3$ rotates the w components of the $\mathbf{R}_A$ vectors back into the 1-2 plane. The sum $\mathbf{R}_T$ is still zero. For $t > t_3$, these transverse components evolve as

$$\begin{aligned} \mathbf{R}_A(t) &= u_A \hat{1} + v_A \hat{2} \\ &= w_A(t_2+)e^{-(t_3-t_2)/T_b}e^{-(t-t_3)/T_2}[\cos(\Delta(t-t_3) - \mathbf{k}_3 \cdot \mathbf{r} - \phi_3)\hat{1} \\ &\quad + \sin[\Delta(t-t_3) - \mathbf{k}_3 \cdot \mathbf{r} - \phi_3]\hat{2}]. \end{aligned} \quad (6.4.9)$$

The radiating polarization is proportional to the integral of $u + iv$ over the detuning. Assuming again a Gaussian distribution of resonant frequencies with width T_2^{*-1}, the polarization becomes

$$\mathbf{P}(t) = \frac{\mu_{ab} T_2^*}{2\sqrt{\pi}} \mathcal{N} e^{-(t_3-t_2)/T_b} e^{-(t-t_3)/T_2} \exp[i(\Omega_{ab}t - \mathbf{k}_3 \cdot \mathbf{r} + \phi_3)]$$

$$\times \int_{-\infty}^{\infty} w_\Delta(t_2+) \exp(-\Delta^2 T_2^{*2}) \exp(i\Delta[t-t_3])\, d\Delta$$

$$= \{-1\} \frac{\mu_{ab}}{4} \mathcal{N} e^{-t_2/T_2} e^{-(t_3-t_2)/T_b} e^{-(t-t_3)/T_2} \exp\left(-\left[\frac{t-t_3-t_2}{2T_2^*}\right]^2\right)$$

$$\times \exp(i[\Omega_{ab}t + (\mathbf{k}_1 - \mathbf{k}_2 - \mathbf{k}_3) \cdot \mathbf{r} - \phi_1 + \phi_2 - \phi_3]). \quad (6.4.10)$$

It is clear from (6.4.10) that the peak of the stimulated echo pulse occurs at $t = t_3 + t_2$, the time in which the transverse components of all of the vectors created by the stimulating pulse precess into a parallel orientation.

Stimulated echoes also occur when the exciting pulse areas are not equal to $\frac{1}{2}\pi$. The characteristics of the echo are unchanged except that the amplitude is reduced by $(\sin \Theta_1 \times \sin \Theta_2 \times \sin \Theta_3)$. Stimulated echoes occur when $\chi T_2^* \ll 1$. In the low-field limit the amplitude scales as χ^3, the line shape becomes slightly more complex, and the radiated frequency equals that of the stimulating laser pulse [4]. As function of time, the peak amplitude of the stimulated echo in a two-level system always varies as (6.4.1) [17, 21].[1] In some systems, the stimulated echo can be produced using exciting pulses of different polarization. The selection rules are similar to those discussed under polarization spectroscopy and degenerate four-wave mixing [17].

The flexibility in propagation direction and polarization inherent in the stimulated echo facilitates the supression of spurious signals due to light scattering [22]. The stimulated echo can also be "stored" and detected via fluorescence. It allows measurement of both the longitudinal and transverse relaxation times in a single experiment. In vapor samples, the nonuniform spatial population distribution tends to vanish as the result of thermal motion, but the spatially uniform component created by copropagating initial pulses still produces an echo. Except for complexity, the main draw-

---

[1] In systems where the excited state decays to a long-lived reservoir, rather than the initial state, the stimulated echo can indicate an anomalous value for T_b. If the decay rate of the excited state is T_b and the reequilibration rate of the lower level is T_a, the proper form for Eq. (6.4.1) is then

$$|P(t_2 + t_3)| \propto \tfrac{1}{2}\{\exp(-(t_3-t_2)/T_b) + \exp(-(t_3-t_2)/T_a)\} \exp(-2t_2/T_2).$$

The long-lived echoes seen in Ref. [17] occur because a hyperfine level of the ground electronic state acts as a reservoir with $T_a \gg T_b$.

6.5 RAMSEY FRINGES

back of the stimulated echo technique is that the echo amplitude is always less than that of the photon echo amplitude and can be difficult to detect.

6.5 RAMSEY FRINGES

The microwave transitions of atoms or molecules in a beam have very narrow linewidths and it has been known since 1948 that such transitions are highly suitable for use as frequency standards. The problem was, however, that the atoms or molecules were moving, and would escape from the cavity containing the microwave radiation in a time short compared to T_2. The resolution obtained in such a single-field microwave spectroscopy experiment was limited by this transit time broadening to a value orders of magnitude worse than that set by transverse relaxation. Ramsey [23] solved this problem by directing the atomic beam through two (or more) separated cavities, each containing radiation with the same frequency and phase. This separated-field method imposed an oscillatory frequency dependence on the transition rate that served to mark the true resonant frequency.

Ramsey's original microwave method is closely related to the concept of storing a first-order FID. An atom moving through the first microwave cavity in a time short compared to T_2 experiences a pulse of radiation as illustrated in Fig. 6.8. When it leaves the cavity, the **R** vector corresponding to the homogeneously broadened transition of the atom has a component in the 1-2 plane,

$$\mathbf{R}_I = \{-1\}\Theta_1 \operatorname{sinc}\Delta t_p \hat{2} + w\hat{3}, \tag{6.5.1}$$

where $\Theta_1 = \int \chi \, dt$ is the pulse area on resonance, t_p is the time required for an atom to move across the cavity,[2] and we have used (2.4.17) in the limit of small coupling.

In the region between cavities, $\mathbf{R}_I$ precesses according to (2.4.20) arriving at the second cavity after a time t_d as

$$\mathbf{R} = \mathbf{R}_{II} = \{-1\}\operatorname{sinc}\Delta t_p\{\Theta_1 e^{-t_d/T_2}(-\sin\Delta t_d \hat{1} + \cos\Delta t_d \hat{2}) + (1 - \tfrac{1}{2}\Theta_1^2)\hat{3}\}, \tag{6.5.2}$$

The field in the second cavity is coherent with the first and alters the **R** vector according to

$$\mathbf{R}'_{II} = \begin{bmatrix} 1 & -\Delta t_p \operatorname{sinc}\Delta t_p & 0 \\ \Delta t_p \operatorname{sinc}\Delta t_p & 1 & \Theta_2 \operatorname{sinc}\Delta t_p \\ 0 & -\Theta_2 \operatorname{sinc}\Delta t_p & 1 \end{bmatrix} \mathbf{R}_{II}. \tag{6.5.3}$$

Positive v components are transformed into positive w components—an increase in the population of the excited level beyond the value due to one cavity. A negative value of v results in a decrease in w. Stimulated emission

---
[2] Assumed short compared to t_d.

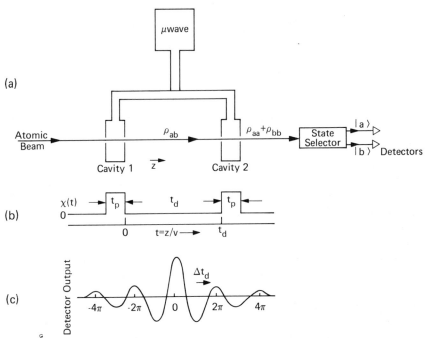

Fig. 6.8 A microwave-frequency separated-oscillatory-fields experiment similar to the original Ramsey fringe technique. (a) An atomic beam transits two cavities oscillating in phase. (b) The time profile of the exciting field. The coherence excited in the first cavity is stored as a population difference in the second. (c) The resulting increase and decrease in the populations of the coupled levels depends on the detuning of the field from resonance as shown at bottom.

in the second cavity causes atoms to return to the lower level. As a function of Δ, the detuning of the microwave frequency from resonance, there is a sinuisoidally varying component of the population difference,

$$\delta w'_{\mathrm{II}}(\Delta) = \{-1\}[-\Theta_1\Theta_2 e^{-t_d/T_2} \mathrm{sinc}^2 \Delta t_p \cos \Delta t_d]. \quad (6.5.4)$$

This frequently dependent population difference can be monitored by a Stern–Gerlach-type state separator or by fluorescence. The oscillatory dependence of this population signal upon the detuning Δ has been called the "Ramsey fringe." Note that the spacing of these fringes can be made smaller than the homogeneous linewidth T_2^{-1} by increasing the distance d between the cavities until $t_d = d/v > T_2^{-1}$, where v is the molecular velocity.

The pattern of the molecular populations $\delta w'$ as a function of Δ resembles the diffraction pattern produced by a pair of slits. The overall width of the diffraction pattern reflects the width of the slits, just as the width of the Ramsey fringe pattern reflects the time the atoms interact with each cavity.

6.5 Ramsey Fringes

The interference fringes in a diffraction pattern reflect the spacing between slits, just as the Ramsey fringe pattern reflects the spacing between cavities. As in the optical case, the addition of more cavities (slits) increases the complexity of the pattern, making the fringe that occurs at $\Delta = 0$ more and more distinct. The cesium frequency standard employs this technique to define the second.

One should also note that the atomic velocities in a beam are not all equal. The correct form of the final Ramsey fringe result is obtained by averaging over the velocities. In this average, the fringes for $\Delta \neq 0$ tend to cancel out, making the central fringe and thus the true resonant frequency even more distinct.

The key to the success of the Ramsey fringe technique in the microwave region was that the two fields could be made to oscillate in phase with one another over the entire regions where the atoms interacted with the field. The absence of spurious phase shifts ensured that the central fringe would appear at $\Delta = 0$ and that the velocity averaging would not cancel it out. In the optical region, it is considerably more difficult to arrange that the fields in two spatially separated regions oscillate with zero phase shift. Moreover, the distances involved are always larger than one wavelength and this can introduce spatially nonuniform phase shifts.

The optical analog of the original Ramsey fringe technique can be applied to the homogeneously broadened transitions produced by the Doppler-free multiquantum absorption techniques. In many of these experiments, a pulsed laser must be used with a pulse length $t_p \ll T_2$. Transit-time broadening is then not as serious a difficulty as is the resolution limit resulting from the transform-limited linewidth of the laser pulse. The two-photon Rabi frequency defined by (2.4.2) and (2.7.8) has no spatial phase dependence for the Doppler-free transitions. The Ramsey fringe technique employed for such transitions by Salour appears in Fig. 6.9 [24]. The atoms are first excited by one pulse of standing wave radiation, and then by a second pulse delayed in time by t_d. A typical experimental trace appears in the inset; Ramsey fringes are clearly visible. The problem is that the delayed pulse is inevitably phase-shifted with respect to the first pulse. Assuming two square laser pulses, with a phase shift ϕ, the population difference after the second pulse has the form

$$\delta w'(\Delta) = \{-1\}[-\Theta_1 \Theta_2 e^{-t_d/T_2} \mathrm{sinc}^2 \Delta t_p \cos(\Delta t_d + \phi)] \qquad (6.5.5)$$

which does not automatically show a central fringe at $\Delta = 0$. Multiple pulses lead to a more complex fringe pattern, but the most distinct fringe still need not correspond to $\Delta = 0$. While this technique may increase the resolution of such a pulsed laser experiment, it does not increase the accuracy with which the atomic frequency can be determined. Ramsey fringe

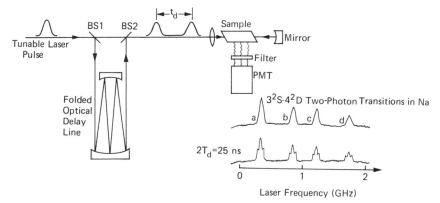

Fig. 6.9 A Doppler-free two-photon Ramsey fringe experiment. The folded optical-delay line stores the second pulse for time t_d. The pulse lengths are much longer than the time needed to reach the retroreflection mirror. The inset shows the Doppler-free two-photon spectrum excited by a single pulse and a trace showing Ramsey fringes. Electronic techniques can emphasize the fringes at the expense of the normal two-photon resonances (from Ref. [24]).

results with similar limitations have been obtained with spatially separated cw laser fields [36].

For the inhomogeneously broadened transitions resolved by saturation spectroscopy, two separated fields will not do [25]. Chebotayev pointed out that the simplest scheme for obtaining Ramsey fringes in saturated absorption must have the three distinct interaction regions diagrammed in Fig. 6.10. The physics of this system is closely related to that of the stimulated echo in the previous system; the field amplitudes, however, are much smaller—typically $\chi T_2 \ll 1$—and the pulse lengths t_p induced by the transit of atoms through the beams are much greater than T_2^*.

In the first interaction region (labeled I), the atoms with frequencies in the laboratory frame within t_p^{-1} of the laser frequency interact with a traveling wave with wave vector $\mathbf{k}_1 = \mathbf{k}$. The Bloch–Feynman vectors for the atoms leaving the interaction have components in the $\hat{1}$-$\hat{2}$ plane similar to, but much smaller than, those given by (6.4.4). In traversing the distance between the first and second interaction regions, the $\mathbf{R}_\Delta$ vectors precess and decay as in (6.4.5).[3]

---

[3] This treatment of Ramsey fringes in an inhomogeneously broadened ensemble differs from the usual presentation which emphasizes that the phase and frequency shifts for the individual molecules result from their motions along linear trajectories between the interaction regions. If the molecular velocity has component v_z along k and component v_d in the direction toward the next interaction region (II), the Doppler detuning $\Delta = k v_z$, and the phase shift resulting from the difference in the atoms axial position in the two interaction regions is $\Phi = k \delta z = k\, dv_z/v_d$. If $v_z \ll v_d$ (the usual Lamb dip case) and if $t_d = d/v_d$, $\Phi = \Delta t_d$. Elementary trigonometry can be used to add this phase shift to ϕ_1 in (6.4.5) and obtain the usual result.

6.5 Ramsey Fringes

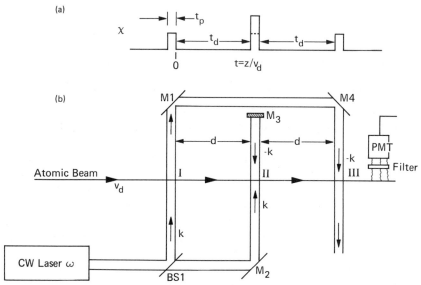

Fig. 6.10 The similarity between Ramsey fringes in saturated absorption and the stimulated echo. (a) The driving fields experienced by a moving atom; (b) schematic of the actual apparatus. The field in interaction regions I and II plays the role of the excitation pulses in the stimulated echo, while the field in region III acts to "store" the stimulated echo so that the photomultiplier can detect the resulting oscillations in the excited-state population.

In the second interaction region (II), the atoms interact with *counter-propagating waves*. These two waves perform the role of the second and third pulses in the stimulated echo sequence with $t_2 = t_3$. The "second" pulse corresponds to interaction with the wave with $\mathbf{k}_2 = \mathbf{k}$, which produces a spatially uniform component of w that varies sinusoidally with Δ as does the first term in (6.4.7). Again, the magnitude of the modulated component is smaller than in the ideal stimulated echo case. The random thermal velocities of the atoms do not scramble the phase of the oscillation in this spatially uniform part of w_Δ.

The third pulse, which is contemporaneous with the second, corresponds to the interaction with the wave in the center region with $\mathbf{k}_3 = -\mathbf{k}$. This wave converts the oscillation in population difference into transverse components of the $\mathbf{R}_\Delta$ vectors with a spatial dependence corresponding to $\mathbf{k}_3$. The ensemble of atoms leaving the second interaction region has $\mathbf{R}_\Delta$ vectors precessing as

$$\mathbf{R}_{\Delta_{\text{II}}} \propto \{-1\} \cos(\Delta t_d + \phi_1 - \phi_2) e^{-t_d/T_2} [-\sin(\Delta(t - t_d) - \mathbf{k} \cdot \mathbf{r} + \phi_3)\hat{1} \\ + \cos(\Delta(t - t_d) - \mathbf{k} \cdot \mathbf{r} + \phi_3)\hat{2}] e^{(-t - t_d)/T_2} + w\hat{3}. \quad (6.5.6)$$

In the third interaction region at $t = 2t_d$, this ensemble interacts with a wave propagating in the $\mathbf{k}_4 = -\mathbf{k}$ direction. The effect of this interaction is to "store" the coherence implied by (6.5.6) as a population difference. Performing a transformation similar to (6.4.6) and collecting terms, we have

$$\delta w_\Delta(2t_d) \propto \{-1\} \cos(\Delta t_d + \phi_1 - \phi_2) \cos(\Delta t_d + \phi_3 - \phi_4) e^{-2t_d/T_2}. \quad (6.5.7)$$

Again this population difference is spatially uniform. The population difference leaving region III may be detected by monitoring the fluorescence from the excited state.

The analysis so far has neglected the fact that only a portion of the Doppler-broadened frequency distribution can interact with the laser beams when $\chi T_2^* \ll 1$ and that different portions interact unless $\Omega_{ab} = \omega$. These issues have been addressed in detail in Chapter 3. To model the effect of transit-time broadening produced by the interactions of a moving atom with counterpropagating laser beams with Gaussian profiles, we assume that $t_p \ll T_2$ and note that the velocity distribution for an atom interacting with waves propagating in the $\mathbf{k}$ direction will be centered at $v_+ = (\Omega_{ab} - \omega)/k$ while those interacting with $-\mathbf{k}$ waves will be centered at $v_- = (\omega - \Omega_{ab})/k = -v_+$. The width of each distribution will be given by the transit-time broadening $\delta v_z \approx t_p^{-1} k^{-1}$. For Gaussian intensity profiles, the velocity distributions then become

$$g_\pm(v_z) = (2\sqrt{\pi}\, \delta v_z)^{-1} e^{-(v_z \pm v_+)^2/\delta v_z^2}. \quad (6.5.8)$$

The proportion of atoms interacting with both waves is then the convolution of the two distribution functions

$$g(v_+) = \int g_+(v_z) g_-(v_z)\, dv_z \sim (t_p/\sqrt{\pi}) e^{-t_p^2 \Delta^2/2} = G'(\Delta), \quad (6.5.9)$$

where the definition of the Doppler detuning $|\Delta| = kv_+$ has been employed. The function $G'(\Delta)$ describes the line shape of the transit-broadened Lamb dip and must multiply the population oscillation term in (6.5.7). The net line-shape function for the Ramsey fringes on the Lamb dip is then

$$\delta w_\Delta \propto \{-1\}(t_p/2\sqrt{2\pi}) e^{-t_p^2 \Delta^2/2} e^{-2t_d/T_2}[\cos(\phi_1 - \phi_2 - \phi_3 + \phi_4) \\ + \cos(2\Delta t_d + \phi_1 - \phi_2 + \phi_3 - \phi_4)]. \quad (6.5.10)$$

Note that δw_Δ at $\Delta = 0$ can be a maximum, minimum, or anything in between depending upon the phase shifts for the four interacting beams. Interferometric techniques, however, can set all of these phase shifts to zero [26].

More complex fringe patterns are obtained when additional interaction regions with counterpropagating waves are added to the apparatus between the initial and final interactions where only a unidirectional wave is necessary.

Also, the time t_d required to move from one beam to the next depends upon the transverse component of the atomic velocity ($t_d = d/v_d$ where d is the physical distance between the three interaction regions). A numerical calculation is necessary to average over the Maxwell–Boltzmann distribution of transverse velocities, unless a monoenergetic beam source is used. Results obtained using just such a source are shown in Fig. 3.18. For a broader distribution of transverse velocities, one would expect that all but the fringe at $\Delta = 0$ would average out.

Finally, at the highest level of resolution, the effect of time dilation in the rest frame of the transversely moving atoms shifts the Ramsey fringe pattern. The time-dilation effect causes the fringe patterns due to different transverse velocity groups to shift with respect to one another. Even if the net phase shift has been set to zero, this time-dilation effect would displace the maximum of the fringe pattern from $\Delta = 0$. A detailed numerical calculation would be required to find the true resonance frequency [27].

6.6 EXPERIMENTAL TECHNIQUES AND RESULTS

The experimental implementation of the optical coherent transient techniques requires an elegance no less than that of the theoretical derivations. To observe spectroscopically useful optical nutations and free-induction decays, one must contrive to couple and decouple the radiation and ensemble in a time much less than T_2. It is not sufficient in practice just to turn the light wave on or off. The nutation and FID are radiated collinearly with the driving field, and thus the detector which converts the optical intensities to electrical signals is also sensitive to changes in the driving field. A 100% modulation of the driving field would introduce electronic artifacts in the detector and its associated circuitry that would often obscure the optical transient signal. Moreover, the radiated optical intensity corresponding to the coherent transient driving polarizations would be weak, scaling as $|P|^2$.

Brewer and others have overcome these difficulties by suddenly changing the detuning $\Delta = \Omega_{ab} - \omega$ by an amount $\delta\omega$ greater than the homogeneous linewidth T_2^{-1} [2, 4]. If $\chi T_2 \leq 1$, such a frequency shift at $t = 0$ decouples the part of an inhomogeneously broadened ensemble previously in resonance, leading to an FID, while coupling a new frequency group which then radiates a nutation. The changes in intensity at the detector result from the interference of the waves radiated by the ensemble with the transmitted light beam. These signals then scale linearly with the transient polarizations as discussed in Section 2.9. The extra sensitivity resulting from this optical heterodyne-detection technique allows the use of well-stabilized but relatively low-powered lasers in frequency-switching transient experiments.

In a frequency-switching experiment, the FID and optical nutation signals appear superimposed upon one another in time. The heterodyne intensity resulting from the FID oscillates at the frequency shift $\delta\omega$ that caused the decoupling, while the optical nutation signal oscillates at χ. Electronic or digital filtering can separate the two transients from electrical signals of the sort in Fig. 6.11.

In the earliest frequency-switching experiments, Shoemaker and Brewer coupled and decoupled the field by shifting the resonant frequency of their sample by means of the linear Stark effect [2, 4]. For molecules with fixed dipole moments, the Stark effects of some angular momentum states are sufficient to shift the resonant frequency by T_2^{-1} when the electric field changes by a few tens of volts per centimeter. Since Stark shifting was necessary to make the transition frequency equal to the laser frequency in the first place, arranging a time-dependent shift was not a major undertaking. Similar effects have been exploited in ruby. Among the remarkable results obtained by Stark switching is the FID trace in Fig. 6.12. In this case, several overlapping transitions were prepared by the initial coupling of the laser and ensemble, but when the Stark shift was applied each *transition shifted in*

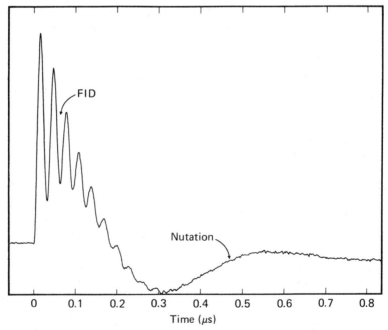

Fig. 6.11 Free-induction decay and optical nutation signals produced in I_2 by frequency switching. The electrical signal due to the FID oscillates at the frequency shift $\delta\omega$, while the nutation oscillates at the Rabi frequency (from Ref. [30]).

6.6 Experimental Techniques and Results

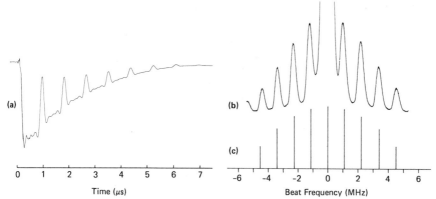

Fig. 6.12 Modulated FID signals observed in $^{13}CH_3F$ by Stark frequency shifting. (a) The actual FID trace; (b) the Fourier transform reveals the frequency splittings; (c) the expected frequency components (from Ref. [2]).

frequency by a different amount [2]. The FID signals beat with each other and with the transmitted laser field producing the complex pattern in Fig. 6.12a. The Fourier transform of this pattern shown in (6.12b) reveals the individual frequency shifts and linewidths.

A more general frequency-shifting technique requires shifting the laser frequency while leaving the resonance unperturbed. Such a frequency shift can be obtained with an acoustooptic modulator which adds in ultrasonic frequency to the incident laser frequency [28]. Such a system is diagrammed in Fig. 6.13a. The acoustooptic modulator, however, deviates the laser beam by an amount proportional to the frequency shift. If the angular deviation is too large, the heterodyne enhancement of the detection sensitivity vanishes. Thus, the available frequency shifts are limited to a few megahertz by the diffraction angle of the laser beam. Such a modulator can switch the driving fields completely on and off as well as switching their frequencies. This feature is useful in photon echo experiments where the stability of a cw source can be an advantage. The field can also be switched on at the time of the expected echo in order to provide a local oscillator for optical heterodyne detection.

Much larger frequency shifts can be obtained (for a limited period of time) by replacing the acoustooptic modulator by an electrooptic phase modulator as in Fig. 6.12b [29]. This device has a voltage-dependent index of refraction, which alters its apparent optical length. Applying a voltage ramp is optically equivalent to moving the laser toward or away from the sample at a velocity v. There thus is an apparent Doppler shift in the exciting frequency so long as the optical length of the modulator is changing. Unfortunately, all voltage

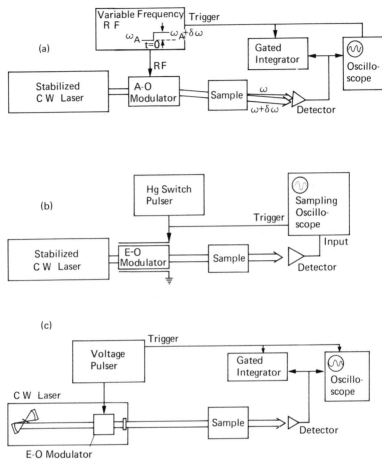

Fig. 6.13 Three coherent transient spectrometers employing cw lasers and optical frequency switching. (a) An acoustooptic modulator varies the frequency and intensity of the field in the sample as the radio-frequency input varies. (b) The extracavity electrooptic modulator can shift the frequency rapidly, but only for a limited time. An intracavity electrooptic modulator shifts the frequency rapidly and can maintain the shifted frequency indefinitely. Data is collected with a boxcar averager or sampling oscilloscope.

ramps must end, and thus the frequency shift cannot be maintained indefinitely. Traveling wave modulators can produce frequency changes within 50 ps, allowing fast relaxation phenomena, such as the first-order FID, to be studied.

Both the preceding extracavity techniques allow the laser to operate undisturbed and thus facilitate sophisticated servo techniques to control the intensity and frequency. A well-stabilized laser is essential to experiments

6.6 Experimental Techniques and Results

involving the long-lived coherence in many samples. In $Pr:LaF_3$ with perturbations due to precession of the fluorine nuclei eliminated, the transverse lifetime of the transition is $T_2 = 66$ μs at $2°K$.

Placement of the electrooptic modulator inside the laser cavity as in (6.12c) allows frequency shifts to be maintained indefinitely [30]. A change of the voltage on such a modulator is equivalent to moving one of the mirrors of the laser cavity, thus altering the frequency of the single mode laser by $\delta\omega = \delta l/\lambda(c/2L)$. Unfortunately, there can be complications due to the dynamics of the gain medium and of the resonator. In organic dyes, the phase memory is very short so the medium is not a problem, but avoiding resonator transients requires that the frequency shift be an integral number of axial mode spacings.

When the homogeneous linewidth is too great to frequency shift entirely out of resonance or when T_2 is longer than the ramp that can be applied to an electrooptical modulator, a variation of frequency switching termed "phase switching" has proved useful [31]. In this scheme, the laser frequency is shifted by $\delta\omega$ for a time $t_s = \pi/\delta\omega$ and then shifted back. The ensemble is not quite decoupled, but if $\chi^2 T_2 T_b \ll 1$, a transient with the characteristics of a free-induction decay will be radiated. After a time χ^{-1}, additional decay processes and nutation signals become apparent, but the interval $0 < t < \chi^{-1}$ can provide enough data to determine T_2^{-1}. Phase switching is closely related to polarization switching, which can be useful for optically isotropic media [32].

Frequency switching can excite echoes so long as $\delta\omega T_2 \gg 1$ and $\delta\omega \gg \chi$ are fulfilled. Large pulse areas are obtained by tailoring the pulse length, which, of course, must be less than T_2. A photon echo produced using acoustooptic amplitude modulation and frequency shifting is shown in Fig. 6.14. As with previous frequency-shifting techniques, the echo signal is modulated at the shift frequency due to the heterodyne beating of the radiated polarization and the transmitted local oscillator laser beam [33]. In multiple-pulse frequency-shift experiments, one must consider the effects resulting from the frequency groups in resonance *between* pulses.

Pulsed lasers such as the N_2 laser pumped dye laser are more easily adapted to the echo techniques. The signals typically detected correspond to the intensities actually radiated by the ensemble. Even though they scale as $|P|^2$, the signals can be strong enough to be detected without heterodyne enhancement because the entire ensemble can be uniformly excited with $\chi T_2^* \gg 1$. The need to shield the detector from the large incident laser beams and from scattered light suggests the use of noncollinear excitation geometries. An apparatus due to Hartmann for producing two- and three-pulse echoes is shown in Fig. 6.5. The timing between pulses is set by the triggering times for independent nitrogen pump lasers [18]. Electrooptical shutters

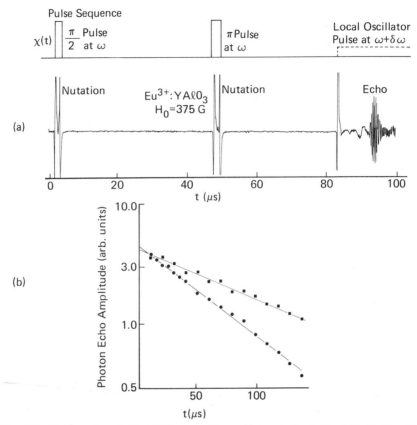

Fig. 6.14 Photon echoes in Eu:YAlO$_3$ excited by apparatus similar to Fig. 6.13. (a) The echo at 100-μs delay is modulated because the local oscillator wave is frequency shifted from that of the exciting pulses. (b) The decay envelope of the optical heterodyne-detected echo amplitude for two magnetic fields: ●, $H_0 = 0$, $T_2 = 58$ μs (5.5 kHz); ■, $H_0 = 375$ G, $T_2 = 104$ μs (3.1 kHz). The inferred homogeneous linewidth of 3.1 kHz implies the highest optical resolution yet achieved (see Ref. [33]).

open only when the desired pulse is expected, and thus partially shield the detectors from scattering, laser fluorescence, and other effects.

Echoes can be studied in the picosecond regime using repetitively pulsed mode-locked lasers with appropriate shutters and detectors. Fayer has employed sum frequency generation in a nonlinear crystal to separate picosecond-stimulated echo phenomena from a nearly overwhelming background level [34]. The idea of this "shutter" is that the crystal generates an output at $\omega_1 + \omega_2$ only when the echo pulse at ω_1 and a local oscillator

6.6 Experimental Techniques and Results

pulse at ω_2 arrive contemporaneously at the crystal from the proper phase-matching directions. A detector placed after the crystal is sensitive only to the sum frequency, which is not otherwise present. By intersecting the echo and local osicllator beams at an angle, the sum frequency direction can be distinguished from the direction of the second harmonic of the local oscillator. This shutter opens and closes in the time of one laser pulse—for Fayer, tens of picoseconds.

Various types of transients have been observed for vibrational and electronic transitions that couple nonlinearly to the incident fields. The appropriate Rabi frequency and dipole moment operators can be defined using the techniques of Chapter 2. Usually the Raman and two-photon transient experiments require two pulsed lasers to couple levels at $\Omega_{rg} = \omega_1 - \omega_2$ or $\Omega_{tg} = \omega_1 + \omega_2$. If one of the frequencies is incident at a later time, the *other* frequency will be produced as the result of a Raman or two-photon first-order FID. Echoes have been observed for a spin-flip Raman transition in CdS by exciting with *two* two frequency pulses and probing with a single frequency at the time of the expected echo [35]. For vibrational transitions in condensed phases, the dephasing is so rapid that picosecond lasers are essential.

In vapors, the relaxation times are longer, and Loy has observed a two-photon optical precession phenomenon in NH_3 using a single-mode pulsed TEA laser and a quasi-cw CO_2 probe laser propagating in opposite directions through the vapor [9]. The optical Stark shift due to the strong TEA laser brought the two-photon transition into resonance, producing a nutation. After the TEA laser pulse, a first-order FID could be observed superimposed upon the probe laser intensity. Liao has observed Doppler-free two-photon electronic transitions in sodium vapor, both with cw lasers by Stark shifting and with pulsed lasers [10]. In the latter case, the first-order FID appeared as the frequency-shifted reflection of the delayed probe pulse. The two-photon FID was identified by the pressure dependence of its decay rate.

It is impossible even to summarize the wide variety of applications and experimental results obtained to date with coherent transient techniques. Dephasing rates have been measured for slowly relaxing transitions in solids such as those giving rise to Figs. 6.2 and 6.14. The dominant dephasing mechanisms have been identified as random spin flips of nuclei near the absorbing ion. Cross sections for decay, dephasing, and frequency shifting have been measured for transitions of Na, I_2, CH_3F, and a number of other gaseous species. Vibrational relaxation in liquids has been studied in order to better understand the liquid state. The beating observed in Figs. 6.6 and 6.11 allows the energy and linewidths of unresolved transitions to be determined.

One- and two-photon optical Ramsey fringe experiments promise to eliminate transit time broadening as a limiting factor in spectroscopic resolution. With multiple intense coherent pulses, even weak transitions can be excited and resolved. Ultimately, it may be possible in this manner to define the meter and second in terms of a single optical frequency transition.

REFERENCES

1. A. Abragam, "Principles of Nuclear Magnetism." Oxford Univ. Press, London and New York, 1961.
2. R. G. Brewer, Coherent optical spectroscopy, *in* "Nonlinear Spectroscopy" (Proc. Int. School. Phys., Enrico Fermi, Course 64) (N. Bloembergen, ed.). North-Holland Publ., Amsterdam, 1977, and references therein. Also, R. G. Brewer, Coherent optical spectroscopy, *in* "Frontiers in Laser Spectroscopy" (R. Balian, S. Haroche, and S. Liberman, eds.), Vol. 1, pp. 343–347. North-Holland Publ., Amsterdam, 1977.
3. T. Mossberg, A. Flusberg, R. Kachru, and S. R. Hartman, *Phys. Rev. Lett.* **34**, 1523 (1977).
4. R. L. Shoemaker, Coherent transient infrared spectroscopy, *in* "Laser and Coherence Spectroscopy" (J. I. Steinfeld, ed.), pp. 197–372. Plenum, New York, 1978, and references therein.
5. J. D. Macomber, "The Dynamics of Spectroscopic Transitions." Wiley New York, 1976.
6. T. W. Mossberg and S. R. Hartman, Diagrammatic representation of photon echoes and other laser-induced ordering processes in gases, *Phys. Rev. A* **23**, 1271–1280 (1981).
7. A. Schenzle and R. G. Brewer, Generalized two-photon theory, *Phys. Rep.* **43**, 457–484 (1978).
8. R. G. Devoe and R. G. Brewer, *Phys. Rev. Lett.* **36**, 959 (1976).
9. M. M. T. Loy, Observation of two-photon optical nutation and free induction decays, *Phys. Rev. Lett.* **36**, 1454–1457 (1976); **39**, 187 (1977).
10. P. F. Liao, N. P. Economou, and R. R. Freeman, *Phys. Rev. Lett.* **39**, 1473 (1977).
11. A. Laubereau and W. Kaiser, *Rev. Mod. Phys.* **50**, 607 (1978); C. H. Lee and D. Ricard, *Appl. Phys. Lett.* **32**, 168 (1978).
12. S. C. Rand, A. Wokaun, R. G. Devoe, and R. G. Brewer, Magic-angle line narrowing in optical spectroscopy, *Phys. Rev. Lett.* **43**, 1868–1871 (1979).
13. A. Schenzle, N. C. Wong, and R. G. Brewer, Theorem on coherent transients, *Phys. Rev. A* **22**, 635–637 (1980).
14. N. A. Kurnit, I. D. Abella, and S. R. Hartman, Observation of a photon echo, *Phys. Rev. Lett.* **13**, 567 (1964).
15. O. Howarth, "Theory of Spectroscopy." Halsted, 1973.
16. P. R. Berman, J. M. Levy, and R. G. Brewer, *Phys. Rev. A* **11**, 1668 (1975).
17. T. Mossberg, A. Flusberg, R. Kachru, and S. R. Hartmann, Total scattering cross section for Na on He measured by stimulated photon echoes, *Phys. Rev. Lett.* **42**, 1665–1669 (1979).
18. Y. C. Chen, K. Chiang, and S. R. Hartman, Spectroscopic and relaxation character of the $^3P_0$–$^3H_4$ Transition in $LaF_3:Pr^{3+}$ measured by photon echoes, *Phys. Rev. B* **21**, 40–47 (1980).
19. S. B. Grossman, A. Schenzle, and R. G. Brewer, *Phys. Rev. Lett.* **38**, 275 (1977).
20. A. H. Zewail, T. E. Orlowski, K. E. Jones, and D. E. Godari, *Chem. Phys. Lett.* **48**, 256 (1977); A. H. Zewail, T. E. Orlowski, and D. R. Dawson, *ibid.* **44**, 379 (1976).
21. J. B. W. Morsink and D. A. Wiersma, Optical coherence storage in spin states, *in* "Laser

References

Spectroscopy IV" (H. Walther and K. W. Rothe, eds.), V. pp. 404–415. Springer-Verlag, Berlin, 1979.

22. M. Fujita, H. Nakatsuka, H. Nakanishi, and M. Matsuoka, Backward echo in two-level systems, *Phys. Rev. Lett.* **42**, 974–977 (1979).
23. N. F. Ramsey, *Phys. Rev.* **78**, 695 (1950).
24. M. M. Salour, Ultrahigh-resolution two-photon spectroscopy in atomic and molecular vapors, *Ann. Phys. (NY)* **111**, 364–503 (1978).
25. V. P. Chebotayev, Coherence in high resolution spectroscopy, *in* "Coherent Nonlinear Optics" (M. S. Feld and V. S. Letokhov, eds.) (Topics in Current Physics 21), pp. 59–109. Springer-Verlag, Berlin, 1980.
26. R. L. Barger, J. C. Bergquist, T. C. English, and D. C. Glaze, Resolution of photon recoil structure of the 6573 Å calcium line in an atomic beam with optical Ramsey fringes, *Appl. Phys. Lett.* **34**, 190–191 (1979).
27. R. L. Barger, Influence of second order Doppler effect on optical Ramsey fringe, *Opt. Lett.* **6**, 145–148 (1981).
28. R. G. Devoe, A. Szabo, S. C. Rand, and R. G. Brewer, Ultraslow optical dephasing of $LaF_3:Pr^{3+}$, *Phys. Rev. Lett.* **42**, 1560–1563 (1979).
29. R. G. Devoe and R. G. Brewer, Subnanosecond optical free induction decay, *Phys. Rev. A* **20**, 2449–2458 (1979).
30. A. Z. Genack and R. G. Brewer, Optical coherent transients by laser frequency switching, *Phys. Rev. A* **17**, 1463–1473 (1978).
31. A. Z. Genack, D. A. Weitz, R. M. Macfarlane, R. M. Shelby, and A. Schenzle, Coherent transients by optical phase switching: Dephasing in $LaCl_3Pr^{3+}$, *Phys. Rev. Lett.* **45**, 438–441 (1980).
32. M. D. Levenson, Coherent optical transients observed by polarization switching, *Chem. Phys. Lett.* **64**, 495–498 (1979).
33. R. M. Shelby and R. M. Macfarlane, Frequency-dependent optical dephasing in the stoichiometric material EuP_5O_{14}, *Phys. Rev. Lett.* **45**, 1098–1101 (1980).
34. D. E. Cooper, R. D. Weiting, and M. D. Fayer, Picosecond time scale optical coherence experiments, *Chem. Phys. Lett.* **67**, 41 (1979).
35. P. Hu, S. Geshwind, and T. M. Jedju, *Phys. Rev. Lett.* **37**, 1357 (1976).
36. This technique has now been extended to Raman transitions. See J. E. Thomas, P. R. Hemmer, S. Ezekiel, C. C. Leiby, Jr., R. H. Picard, and C. R. Willis, Observation of Ramsey fringes using a stimulated resonant Raman transition in a sodium atomic beam, *Phys. Rev. Lett.* **48**, 867–870 (1982).

Chapter 7
NONLINEAR SOURCES FOR LINEAR AND NONLINEAR SPECTROSCOPY

It seems to be a fact of nature that the convenient tunable laser sources operate in the visible or in the spectral regions adjacent to the visible. The electromagnetic spectrum, however, continues to infinitely high and infinitely low frequencies. Most of the interesting spectroscopy thus occurs beyond the range spanned by the dye laser, the F-center laser, the parametric oscillator, and the other familiar tunable lasers. Nonlinear optical effects can be used to extend the tuning ranges of these sources by combining optical frequencies or the frequencies of material transitions with the tunable laser frequency. Such nonlinear sources extend tunability into portions of the extreme ultraviolet (XUV), vacuum ultraviolet (VUV), ultraviolet (UV), middle infrared (MIR), and far infrared (FIR). A few of the most useful nonlinear frequency generation techniques will be summarized in this chapter.

7.1 SECOND HARMONIC AND SUM FREQUENCY GENERATION

By far the most familiar nonlinear optical device is the second harmonic generation crystal used to double the frequency of commercial Nd:YAG lasers and thus produce radiation useful for pumping dye lasers. The second harmonic generation process relies upon the nonvanishing second-order nonlinear susceptibility tensor $\chi_{ijk}^{(2)}(\omega_3, \omega_1, \omega_2)$ in crystals that lack a center-of-inversion symmetry. The driving polarization in the notation of Chapter 2

7.1 Second Harmonic and Sum Frequency Generation

is

$$P_i^{(2)}(\omega_3) = \chi_{ijk}^{(2)}(\omega_3, \omega_1, \omega_2) E_j(\omega_1) E_k(\omega_2) e^{-i\omega_3 t}, \tag{7.1.1}$$

where $\omega_3 = \omega_1 + \omega_2$, and $\omega_1 = \omega_2$ for second harmonic generation. The beam emerges from the crystal collinearly with the fundamentals. Entire texts have been written on the physics and engineering of devices employing this nonlinearity [1, 2]. For the purposes of spectroscopy it is sufficient to note that materials exist capable of generating the second harmonic of an input down to a wavelength of ~ 2400 Å, that sum frequency generation works down to a wavelength of ~ 2000 Å, and that difference frequency generation does not work very well at all [3].

Relatively large harmonic generation efficiencies are possible because the birefringence of certain crystals can be used to counteract the dispersion of the index of refraction and thus fulfill the wave-vector matching condition

$$\Delta k = |\mathbf{k}(\omega_3) - \mathbf{k}(\omega_1) - \mathbf{k}(\omega_2)| \to 0. \tag{7.1.2}$$

The corresponding condition for the index of refraction is

$$n(\omega_3, \theta) = \tfrac{1}{2}[n(\omega_1, \theta) + n(\omega_2, \theta)]. \tag{7.1.3}$$

The output power of a crystal of length l has the familiar

$$\mathscr{P}^{(2)} \propto |\sin(\Delta k l/2)/(\Delta k l/2)|^2 l^2 \tag{7.1.4}$$

dependence upon Δk and l. In the ultraviolet, the birefringence of many crystals becomes insufficient to counteract the index dispersion, thus making phase matching impossible at wavelengths where $\chi^{(2)}$ and the crystal absorption are still acceptable.

In a uniaxial crystal, the index of refraction for the extraordinary ray—that is, the ray polarized with a projection of the $\mathbf{E}$ vector along the optical axis $\hat{c}$—depends upon the angle between $\mathbf{k}$ and $\hat{c}$ (see Fig. 7.1) [2]

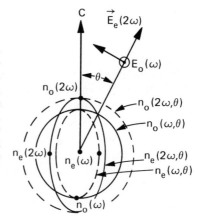

Fig. 7.1 Index of reflection surfaces for a negative uniaxial crystal. Dispersion shifts the indices for the second harmonic from those of the fundamental. The two surfaces relevant for type-I index matching are drawn as solid lines, and the phase matching direction is indicated. The polarization directions for the second harmonic (extraordinary wave) and fundamental (ordinary wave) are shown as is the optical or "c" axis of the crystal. The eccentricities are exaggerated (from Ref. [2]).

$$\frac{1}{n_e^2(\omega, \theta)} = \frac{\cos^2 \theta}{n_0^2(\omega)} + \frac{\sin^2 \theta}{n_e^2(\omega)}. \quad (7.1.5)$$

The index for the ordinary ray—the ray polarized perpendicular to $\hat{c}$—is $n_0(\omega)$ independent of angle. Phase matching is achieved in most such crystals by adjusting θ until the condition in (7.1.3) is fulfilled. Beam divergence is inconsistent with perfect phase matching; such divergence introduces a range of angles θ.

In type-I phase matching for second harmonic generation, the fundamental propagates as an ordinary wave, while the second harmonic propagates as an extraordinary wave. Phase matching is achieved by adjusting the angle θ until

$$n_e(2\omega, \theta) = n_0(\omega). \quad (7.1.6)$$

For sum generation, the corresponding type-I phase matching condition is

$$n_e(\omega_1 + \omega_2, \theta) = \tfrac{1}{2}[n_0(\omega_1) + n_0(\omega_2)]. \quad (7.1.7)$$

The output of a type-I crystal is polarized perpendicular to the inputs, and thus can be easily separated. The angular precision required to achieve phase matching is rather high, and thus the input beams must be well collimated. At $\theta = 90°$, a special condition is reached where angular sensitivity is minimized and temperature tuning must be used to achieve phase matching. This 90° phase-matching condition often gives the best efficiency, especially for low-power beams, but is often cumbersome when the input frequency must be varied.

In type-II phase matching, the fundamental wave is split into two components, one polarized with a component parallel to $\hat{c}$ and one perpendicular. The harmonic or sum frequency again propagates as an extraordinary wave. The index matching condition for sum generation is

$$n_e(\omega_1 + \omega_2, \theta) = \tfrac{1}{2}[n_e(\omega_1, \theta) + n_0(\omega_2)] \quad (7.1.8)$$

which is less critical than the type-I condition. Thus, type-II phase matching is more tolerant of beam divergence and other imperfections. In type-II SHG, the fundamental exits the crystal with essentially random polarization.

Since the indices of refraction are frequency dependent, Eqs. (7.1.2) and (7.1.8) require that the phase-matching angle vary with fundamental frequency. Crystal temperature also affects phase matching, and the fundamental and harmonic beams are often strong enough to alter the temperature. Thus, a tunable laser system employing harmonic generation must incorporate some servo system to optimize the crystal angle. One such system appears in Fig. 7.2 [4]. Portions of the fundamental beam that has been

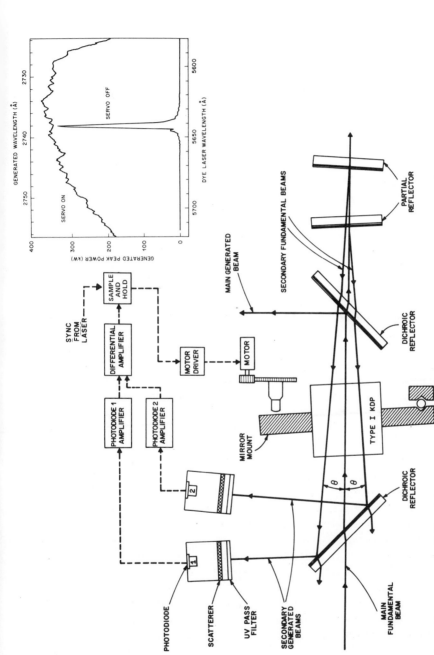

Fig. 7.2. Servo system to scan the matching angle of a harmonic generation crystal as dye laser wavelength varies. Dichroic reflector separates the harmonic from the fundamental, and two fundamental beams are reflected back through the crystal, one on either side of the main beam. Balancing the second harmonic intensities on the two photodiodes ensures that the main beam propagates at optimum phase matching. Inset shows harmonic intensity with and without the servo. Increased tuning range is evident in figure. The servo also helps to compensate for thermal drifts due to absorption in the crystal (from Ref. [4]).

transmitted through the crystal are reflected back through the crystal with small angular deviations from the fundamental. Photodiodes sense the generation efficiency for these reflected beams, and a servo adjusts the crystal angle to equalize the efficiencies. In this way, the phase matching for the main beam propagating along the bisector of the two reflected beams is optimized. The inset in Fig. 7.2 shows the improved tuning range that results. The peak power conversion efficiency was 20%. For a larger tuning range, a variety of harmonic generator crystals with different angles between the $\hat{c}$ axis and the polished crystal faces must be used.

7.2 THIRD- AND HIGHER-ORDER SUM AND HARMONIC GENERATION

The birefringence of crystals is insufficient to counter the large dispersion between the fundamental and its third harmonic. While third, fourth, and fifth harmonics can be generated down to roughly 2000 Å by successively summing lower harmonics using nonlinear crystals, another strategy is required to generate radiation in the vacuum ultraviolet and beyond. One such strategy utilizes the multiphoton resonant nonlinear mixing processes in atoms described in Section 5.4. Related schemes generate even shorter wavelengths at frequencies well away from such resonances. While the nonlinear susceptibilities for these processes can be calculated by the techniques of Chapter 2, the experimental generation efficiencies fall well below the theoretical prediction as the result of such competing effects as ionization and the population of excited states. Wave-vector matching is as serious an issue as in the previous case, and a number of clever strategies have been developed to optimize phase matching.

The energy-level scheme for a prototypical third-order sum-frequency generation experiment appears in Fig. 5.19. The energy of two input photons approaches a two-photon absorption in the medium. This sort of resonance enhances the sum generation amplitude without attenuating the input or output frequencies. A tunable laser adds its frequency to that of the resonant sum producing an output at $\omega_3 = \omega_0 + \omega_1 + \omega_2$. All the input frequencies would be equal in a third harmonic generation experiment [5].

When the output frequency lies above the main resonance line and the input frequencies lie below, the atomic medium producing the harmonic generation can have an index of refraction at the sum frequency that is less than at the fundamentals. Correct phase matching can then be achieved by adding a normally dispersive gas to the medium until

$$3n_{\text{mix}}(\omega_0 + \omega_1 + \omega_2) = n_{\text{mix}}(\omega_0) + n_{\text{mix}}(\omega_1) + n_{\text{mix}}(\omega_2) \quad (7.2.1)$$

7.2 Third- and Higher-Order Sum and Harmonic Generation

is fulfilled. Away from one-photon resonances, the indices of refraction of a mixture can be calculated from Sellmeier's equation by adding the contributions of species X to that of species Y,

$$n(\omega)_{\text{mix}} = 1 + 2\pi c^2 r_e \left\{ N_x \sum_i \frac{f_{xi}}{\Omega_{xi}^2 - \omega^2} + N_y \sum_i \frac{f_{yi}}{\Omega_{yi}^2 - \omega^2} \right\}, \quad (7.2.2)$$

where f_{xi} and f_{yi} are the oscillator strengths of the transitions in species X and Y having resonant frequency Ω_{xi}, Ω_{yi}, respectively, N_x and N_y are the number densities (atoms/cm^3) and $r_e = 2.818 \times 10^{-13}$ cm.

For third harmonic generation in strontium vapor at the $5s^2 \to 5s5d$ two-photon resonance, optimum phase matching is obtained with 53 atoms of xenon for each atom of strontium. The maximum conversion efficiency to 1919 Å is of order 0.01% [6].

Optimum phase matching need not give maximum sum generation in these systems. For one thing, in the usual tight focusing geometry, there is a phase shift in the fundamental from one side of focus to the other. Perfect phase matching results in a destructive interference between the waves radiated on each side of the beam waist. Collisional effects due to the xenon or other index matching gas tend to populate excited states, destroy coherence, and broaden absorptions. Such phenomena reduce the output signal beyond a certain point by more than improved phase matching can enhance it.

Bjorklund has shown that any two-photon resonant four-wave mixing (third-order sum generation) process can be phase matched if there are three input frequencies. Such a process has one free parameter once the resonance and output frequency are defined, and that frequency can be adjusted to fulfill (7.2.1). This phase-matching scheme is independent of atomic density and does not require any particular mixing ratio [7].

Angular momentum selection rules can be employed to discourage unwanted third harmonic signals in certain sum generation experiments. The third harmonic photon can carry away at most one unit of angular momentum, while the inputs can carry three. Thus, with circularly polarized fundamentals, one of the quanta must have the opposite sense of polarization to the other two for a finite signal to be generated [6].

Third-order sum-generation experiments are usually undertaken in heat pipe ovens with densities of the nonlinear species approaching 10^{15} atoms/cc. The buffer gas used to protect the windows from reactive vapor can act to enhance phase matching if the total gas pressure is above the vapor pressure of the nonlinear species at the operating temperature. At higher temperatures, the metal vapor used as a nonlinear medium drives the buffer gas away from the focal region, necessitating the Bjorklund phase-matching scheme.

Third-order mixing has been used successfully to generate radiation down to roughly 900 Å, but attempts to cascade such processes to shorter wavelength have met with little success. Six-, eight-, and ten-wave mixing mediated by the fifth-, seventh-, and ninth-order nonlinear susceptibility have produced detectable radiation down to 380 Å. In these experiments, the opacity of the nonlinear medium itself was a major consideration. The best results were obtained using helium or neon. Since no material windows are available in this wavelength range, the generated radiation must exit the medium through a small hole in the pressure chamber into a differentially pumped vacuum monochromator. Reintjes *et al.* found it advisable to focus the incoming radiation tightly at this exit aperture [8]. A shock wave developed as the result of the supersonic flow of the gas through the aperture, and it was possible to place the beam waist so that only one half of the focus was in a nonlinear medium while the remainder was in the vacuum of the monochromator. In this way, destructive interference due to the phase shift at focus was eliminated.

The input pulses to the XUV generator had energies of 30 mJ in a pulse length of 20 ps, mostly at 2661 Å. Seventh harmonic generation produced an output detectable on the solar blind photomultiplier at 380 Å, while fifth harmonic generation gave 532 Å. A number of additional wavelengths could be generated by substituting another wavelength for one of the 2661 Å photons. Using the fundamental and second harmonic of their Nd:YAG laser, Reintjes *et al.* observed six different wavelengths due to such high-order mixing processes. The greatest conversion efficiency observed was only 0.001%, but in the XUV wavelength region, short pulses of intense narrow-band coherent radiation can be useful even at the nanojoule level. It would be relatively easy to add a tunable input to such a source to produce truly tunable XUV radiation.

7.3 RAMAN SHIFTING

Stimulated Raman oscillation has been used since the early sixties to downshift the frequency of a laser by an amount equal to the Raman mode of a molecule [9]. For most molecules, the Raman mode with the greatest gain has a frequency of <1000 cm^{-1}, resulting in a shift less than the tuning range of a dye laser. Modern Raman shifting techniques employ vibrational modes in hydrogen, deuterium, methane, and a few other gases which provide shifts of several thousand inverse centimeters (4155 cm^{-1} for H_2) [10]. The nonlinear interactions in such a gas cell are similar to those discussed in Chapter 4. The output frequency from one process can act as the input for the next, ultimately shifting visible wavelengths into the vacuum ultraviolet or middle infrared. The efficiency can be remarkably large, commonly 70%

7.3 Raman Shifting

of the pump radiation is shifted to other frequencies. The correct description of such high amplifications requires the coupled-mode formalism of Ref. [9].

A typical hydrogen Raman shifting apparatus is shown in Fig. 7.3. The gas is contained at high pressure (up to 30 atm) in a cell of roughly 50-cm length. The input radiation is loosely focused into the center of the cell by a lens with roughly 30-cm focal length. A similar lens transparent at the desired output frequency recollimates the radiation. A prism separates the many different wavelengths generated directing the desired wavelength into the experiment.

The threshold for stimulated Raman oscillation in such a system is roughly 1 MW and is independent of the exact geometry so long as the entire focal region is contained in the Raman active medium. Once the threshold for oscillation has been exceeded—and it can be greatly exceeded with present pulsed lasers—the Stokes-shifted frequency builds in intensity until it begins to deplete the pump radiation. There is no self-focusing or self-phase modulation in such a system, and the linewidth of the Stokes shifted beam equals the width of the laser input plus a fraction of the width of the Raman resonance. The threshold does not depend upon the input laser linewidth.

Several things happen simultaneously once the field at the first Stokes frequency has built to an appreciable value. Resonant CARS-type interactions in the hydrogen mix the input and Stokes to generate the first anti-Stokes frequency $\omega + \Omega_R$; CSRS-type processes and stimulated Raman gain with the first Stokes as input generate the second Stokes frequency $\omega - 2\Omega_R$. These anti-Stokes and higher-order Stokes components act as inputs in cascades generating up to four orders of Stokes and nine orders of anti-Stokes radiation, each shifted from the laser frequency by precisely the Raman frequency of the molecule. The tuning range of the various Raman

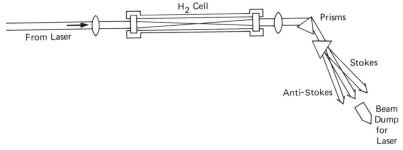

Fig. 7.3 Shifting a laser frequency with a hydrogen Raman cell. The laser beam is focused loosely into the multiatmosphere gas cell where stimulated Raman oscillation and four-wave mixing generate a number of Stokes and anti-Stokes shifted frequencies. Another lens collimates the beam, which is then dispersed with prisms. The unwanted radiation is directed into a beam dump, while the desired wavelengths illuminate the experiment.

shifted frequencies when the input scans over typical dye gain bands is shown in Fig. 7.4. Performance can be improved in the ultraviolet by pumping the Raman cell with the second harmonic of the dye laser [11].

The Stokes-shifted components tend to propagate as diffraction-limited beams collinear with the pump. The successive anti-Stokes components propagate as diverging cones of radiation as a result of the phase matching condition for CARS-type mixing.

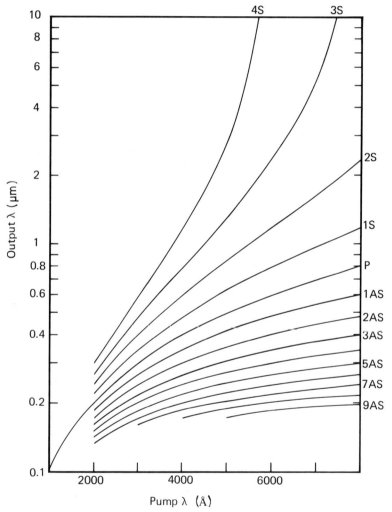

Fig. 7.4 Output wavelengths for the hydrogen Raman shifter cell as a function of the input wavelength. The Nth Stokes and anti-Stokes shifted components are labeled NS and NAS, respectively, while the pump wavelength is labeled P.

7.3 Raman Shifting

The output at a particular frequency can be maximized by varying the pressure, pump power, and (to some extent) the focusing geometry. Typical pulse energies for the various frequency components appear in Fig. 7.5. Maxwell's equations do not permit the longer-wavelength components to be as efficiently driven by the CSRS mixing process as are the corresponding anti-Stokes frequencies. Still, Cahen et al. have reported 50-kW pulses of 16-μm radiation generated by frequency shifting a ruby pumped dye laser in hydrogen [12].

If the input radiation is polarized circularly or elliptically, stimulated Raman oscillation will occur at a variety of rotational frequences and at combinations of rotational and vibrational frequencies. In H_2, the dominant rotational Raman shift is 588 cm^{-1}, much less than the vibrational frequency. Again, cascade processes occur at sufficient intensity. The simplicity and high efficiency of the hydrogen Raman shifter makes it a useful accessory for a high power tunable laser. The main difficulty in using it is that inevitable

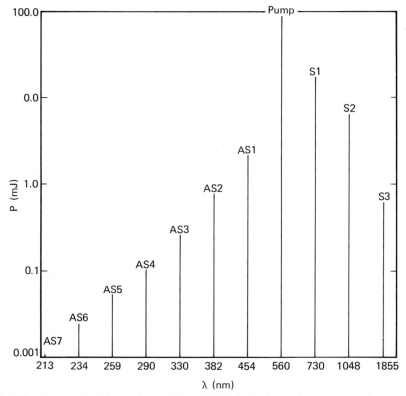

Fig. 7.5 Pump and hydrogen Raman shifter output energies for a pulsed dye laser operating at 560 nm and 85 mJ in a 5-ns pulse. The Nth Stokes and anti-Stokes shifted components are labeled SN and ASN, respectively.

pulse to pulse variations in the transverse profile of the outputs and in the partitioning of energy among the frequency components appears as noise in the final data.

7.4 SPONTANEOUS XUV ANTI-STOKES

A tunable source does not have to be coherent to be useful in spectroscopy. Harris *et al.* have developed a class of XUV sources based upon spontaneous anti-Stokes scattering of tunable radiation from a metastable excited state populated by an electrical discharge [13]. A typical state of this sort is the 1s2s ^{1}S of helium at 166,277 cm^{-1} which can decay only by spontaneous two-photon emission. Radiation from a tunable visible laser can be scattered by atoms initially in this state producing anti-Stokes shifted XUV radiation at $\Omega + \omega$. For common laser dyes, this photon lies between 183,000 and 187,000 cm^{-1}. The linewidth of this radiation is equal to the sum of the laser linewidth and the Doppler width of the excited level—roughly 1.2 cm^{-1} for the present case.[1]

A typical experiment is employing this XUV source appears in Fig. 7.6. Helium is excited to its metastable state by a hollow cathode discharge, and a tunable laser beam with an energy of 50 mJ/pulse is incident. One photon in 10^{14} is spontaneously anti-Stokes scattered in the backward direction. Even so, the source produces 2000 XUV photons/sec with a spectral brightness far above conventional sources.

The XUV photons propagate into the heated sample region where they may be absorbed by a metal vapor confined in that region by the helium buffer gas. The quanta transmitted through the sample are detected by a solar-blind electron multiplier shielded with a 1500 Å aluminum filter. The

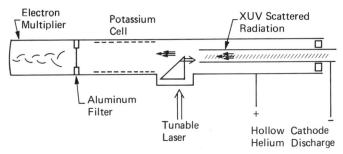

Fig. 7.6 An extreme ultraviolet (XUV) spectroscopy experiment employing an incoherent anti-Stokes scattering source. Tunable radiation is scattered from helium atoms discharge excited into the metastable ^{1}S state. XUV radiation transmitted through the potassium sample is detected by the electron multiplier (from Ref. [13]).

[1] Laser action from transitions of this sort has now been reported by J. C. White and D. Henderson, who prepared a population inversion photochemically [16].

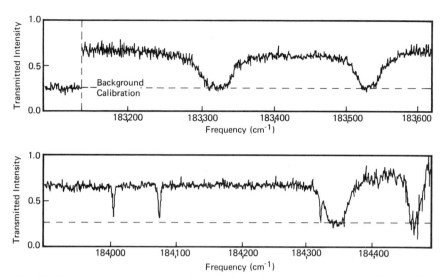

Fig. 7.7 Transmission spectrum of potassium obtained using the apparatus in Fig. 7.6. The broader peaks had previously been observed using conventional XUV sources; the narrow lines are new and are attributed to very long-lived autoionizing states. The fact that the XUV signal falls to the level of the incoherent background from the helium plasma at the maxima of the absorption lines indicates that no laser radiation is reaching the detector and causing multiquantum photoemission (from Ref. [13]).

backscattering geometry and filter are required to separate the true XUV photoelectron signals from a background due to multiquantum photoemission that would have been produced had the dye laser radiation reached the photodetector.

In the initial experiments, Harris *et al.* studied the autoionizing states of potassium. Part of one of their spectra appears in Fig. 7.7. The dotted line indicates the photoelectron signal level when the laser was off. Since the absorption features fall to this level, one may infer that no other frequency components are produced by the nonlinear interaction. The broad absorption features had previously been detected. The narrow features are, however, new and interesting. The widths of these features are less than 1 cm^{-1} larger than the spectral width of the source. This narrow width implies that the states responsible are very long-lived—probably identifiable as "autoionizing" states forbidden by symmetry to ionize.

7.5 INFRARED SPECTROPHOTOGRAPHY

In infrared spectroscopy, it is not always sufficient to generate intense coherent radiation. It is also necessary to detect that radiation, occasionally with a time resolution superior to common infrared detectors. A method combining generation of coherent infrared, efficient detection, and time resolution has been developed by Bethune *et al.* and given the name infrared spectrophotography [14].

In the first stage of their process (illustrated in Fig. 7.8a) a broad-band beam of infrared is generated by stimulated (Stokes) electronic Raman scattering in alkali metal vapor. The radiation incident on the metal vapor is amplified superfluorescence generated by a mirrorless superfluorescent dye

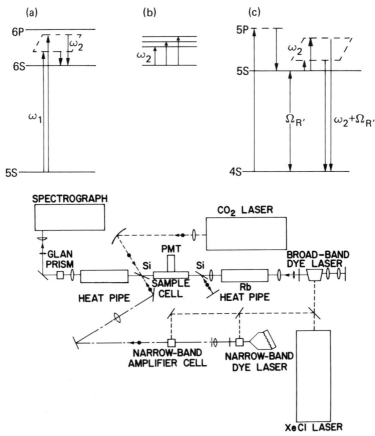

Fig. 7.8 Time-resolved infrared spectrophotography. The experimental apparatus is diagrammed at the bottom of the figure while the relevant level diagrams appear above. (a) The broad-band dye laser output is Raman shifted in the first heat pipe. (b) The resulting infrared is partially absorbed in the sample cell and then (c) shifted back into the visible by the four-wave interaction. This shifting is accomplished in the second heat pipe which is pumped by the narrow-band dye laser. A CO_2 laser is employed to initiate a chemical reaction in the sample cell, while silicon plates are used as infrared transmitting visible reflecting filters (from Ref. [15]).

7.5 Infrared Spectrophotography

cell. Its width of 1000 cm^{-1} is transferred to the Stokes frequency $\omega_2 = \omega_1 - \Omega_R$ in the 3- to 4-μm region.

This infrared radiation passes through a molecular sample, and certain frequency components are absorbed by vibrations in the "fingerprint" region of the molecular spectrum (see Fig. 7.8b). The transmitted infrared is incident upon a second alkali metal vapor cell in which it stimulates the four-wave mixing process diagrammed in Fig. 7.8c. The net effect of the four-wave mixing is to generate a visible photon at frequency $\omega_2 + \Omega_R$ for every frequency component ω_2 of the transmitted infrared. In this way, the infrared absorption spectrum of the molecule is transferred into the visible where it can be dispersed spectrographically, and recorded photographically and photoelectrically.

The infrared spectrophotography process requires a broad-band dye laser for the first stage, synchronized with a narrow-band laser for the second stage (see Fig. 7.8). Both can be pumped with a single Nd:YAG or excimer source. Because the pulse lengths for such lasers are measured in nanoseconds, the entire infrared spectrum can be recorded with nanosecond time resolution. Two heat-pipe ovens, a spectrograph, and a sample cell complete the apparatus.

A typical experimental result appears in Fig. 7.9. The spectrum of CH_3NC is shown as it evolves on a microsecond time scale after the isometrization reaction to CH_3CN is triggered thermally. The v_1 absorption band clearly

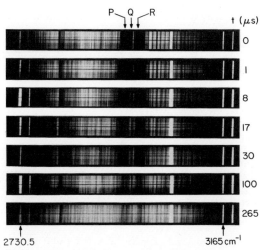

Fig. 7.9 Changes in the spectra of CH_3NC as it thermally explodes (isomerizes). The times are indicated at right, while the spectral features and calibration lines at various times after the application of a CO_2 laser pulse are labeled at top. The broadening of the P and R branches during the first 30 μs indicates rapid heating (from Ref. [14]).

weakens as time goes on, but more significantly the band contours of the P and R branches first broaden and then narrow. The separation between the maxima of the contours is proportional to the square root of the absolute temperature, which increases to 673°K 30 μs after the reaction is initiated.

The evolution of other chemical reactions can also be probed on a nanosecond time scale with this technique. A few dozen laser shots suffice to record the entire "fingerprint" region with adequate sensitivity and signal-to-noise ratio. Spectra taken with fewer shots are degraded by noise analogous to the familiar laser speckle. For repetitive phenomena, this need not be a problem.

REFERENCES

1. F. Zernike and J. E. Midwinter, "Applied Nonlinear Optics." Wiley, New York, 1973.
2. A. Yariv, "Quantum Electronics," 2nd ed., Chapters 16 and 17. Wiley, New York, 1975.
3. Y. R. Shen, Far infrared generation by optical mixing, *Proc. Quant. Electron* **4**, 207–232 (1977).
4. G. C. Bjorklund and R. H. Storz, Servo tuning and stabilization of nonlinear optical crystals, *IEEE J. Quant. Electron.* **QE-15**, 228–232 (1979).
5. J. A. Armstrong and J. J. Wynne, The nonlinear optics of autoionizing resonances, *in* "Nonlinear Spectroscopy" (N. Bloembergen, ed.), pp. 152–169. North-Holland Publ., Amsterdam; R. T. Hodgson, P. P. Sorokin, and J. J. Wynne, *Phys. Rev. Lett.* **32**, 343 (1974).
6. P. P. Sorokin, J. J. Wynne, J. A. Armstrong, and R. T. Hodgson, Resonantly enhanced, nonlinear generation of tunable, coherent, vacuum ultraviolet (VUV) light in atomic vapors, *Ann. NY Acad. Sci.* **267**, 30–50 (1979).
7. G. C. Bjorklund, J. E. Bjorkholm, P. F. Liao, and R. H. Storz, Phase matching of two-photon resonant four-wave mixing processes in alkali metal vapors, *Appl. Phys. Lett.* **39**, 729–732 (1976).
8. J. Reintjes, C. Y. She, and R. C. Eckhardt, Generation of coherent radiation by fifth and seventh order frequency conversion in rare gases, *IEEE J. Quant. Electron.* **QE-14**, 581–598 (1978).
9. N. Bloembergen and Y. R. Shen, *Phys. Rev. Lett.* **12**, 504 (1964); see also N. Bloembergen, "Nonlinear Optics." Benjamin, New York, 1965.
10. V. Wilke and W. Schmidt, Tunable coherent radiation source covering a spectral range from 185 to 880 nm, *Appl. Phys.* **18**, 177–181 (1979).
11. G. C. Bjorklund, private communication.
12. J. Cahen, M. Clerc, and P. Rigny, A coherent light source, widely tunable down to 16 μm by stimulated Raman scattering, *Opt. Commun.* **21**, 387–390 (1977).
13. J. E. Rothenberg, J. F. Young, and S. E. Harris, High resolution spectroscopy of potassium using anti-Stokes radiation, *Opt. Lett.* **6**, 363–365 (1981).
14. P. S. Bethune, J. R. Lankard, M. M. T. Loy, J. Ors, and P. P. Sorokin, *Chem. Phys. Lett.* **57**, 479 (1978).
15. P. N. Avouris, D. S. Bethune, J. R. Lancard, J. A. Ors, and P. P. Sorokin: Time resolved infrared spectrophotography—Study of laser-initiated explosions in HN_3, *J. Chem. Phys.* **74**, 2304–2312 (1981).
16. J. C. White, private communication, also proceedings of I.Q.E.C. XII (To be published).

Appendix
SYMBOL GLOSSARY–INDEX

$\|a\rangle$	state a wave function, 29
A_{ab}	spontaneous emission rate, 3, 69
$A(\Delta\omega)$	isotropic nuclear response at frequency difference $\Delta\omega$, 125, 150
$\alpha(t)$	molecular susceptibility, 3
$\alpha\beta\gamma\delta$	subscript permutation of dipole operators, 59
$\alpha_{\alpha\beta}^R(-\omega_1,\omega_2)$, $\alpha_{\delta\gamma}^R$	Raman susceptibility tensor, 4, 53, 116, 125, 130, 150
α_{eff}^R	effective Raman tensor for given polarization, 128, 135, 150
$\alpha_{\alpha\beta}^T$	two-photon susceptibility tensor, 186
$\partial\alpha/\partial Q$	Raman susceptibility for mode Q, 3, 4, 18
$\|b\rangle$	state b wave function, 29
B	stimulated emission/absorption coefficient, 2
$B(\Delta\omega)$	anisotropic nuclear response at frequency difference $\Delta\omega$, 125, 150
$\boldsymbol{\beta}, \boldsymbol{\beta}(t)$	pseudofield vector, 37, 38, 208
$\beta = \|\boldsymbol{\beta}\|$	Rabi flopping frequency, 37, 38, 199
$\boldsymbol{\beta}'$	pseudofield in nonrotating frame, 36

$\chi(t), \chi(r), \chi, \chi_0$	Rabi frequency, 34, 177, 195, 197, 199, 218
χ_-, χ_+	Rabi frequency for $-$ and $+$ waves, 67
$\chi^T_{\alpha\beta\gamma\delta}, \chi^T$	two-photon resonant nonlinear susceptibility, 125, 186
$\chi^Q_{\alpha\beta\gamma\delta}(-\omega_S,\omega_0,\omega_1,-\omega_2)$	$\chi^{(3)}$ due to mode Q, 118
$\chi_{ij}, \vec{\vec{\chi}}$	linear susceptibility, 2, 4, 46
$\chi^{(2)}_{ijk}$	second-order nonlinear susceptibility, 2, 4, 226
$\chi^{(3)}, \chi^{(3)}_{ijkl}$	third-order nonlinear susceptibility, 4, 46, 67, 77, 81, 122, 130, 149, 189
χ^{NR}	nonresonant $\chi^{(3)}$, 118, 135, 150
χ^{NR}_{eff}	effective nonresonant $\chi^{(3)}$ in given polarization, 128, 135, 150
χ^R	Raman resonant $\chi^{(3)}$, 118, 135
$\chi_\omega(t)$	Rabi frequency for ω frequency wave, 208
χ^B	electronic background $\chi^{(3)}$, 125
d	intercavity distance, 212
D	combinatorial permutation factor, 47, 128, 135
Δ	detuning, 37, 67, 201, 209, 215, 217
ΔI	intensity change, 62, 68
ΔE_u	shift of energy level u, 167
Δk	wave vector mismatch, 131, 190, 227
$\Delta\kappa$	change in absorption constant, 68
δ_L	phase shift, 173
δl	change in optical length, 221
Δn	change in index of refraction, 69
$\Delta\nu$	detection bandwidth, 157
$\Delta\omega$	laser frequency difference, 125, 150
$\delta\omega$	laser frequency shift, 218
$\Delta\Omega$	energy level splitting, 205
$\Delta\Omega_{ab}$	Doppler width of $|a>-|b>$ transition, 82
$<\delta\rho^2_L>$	mean square noise response, 157
δu	mean change in axial velocity, 205

Appendix: Symbol Glossary–Index

δv_z	velocity width, 216
$\mathbf{E}_+, E_+$	electric field for + wave, 66, 166
$\mathbf{E}_-, E_-$	electric field for − wave, 66, 77, 164
E_I	ionization energy, 172
$E_a^\pm, E_b^\pm$	energy of Autler–Townes states, 70
E_b, E_g	energy of level b or g, 29, 164
$\mathbf{E}_{LO}$	local oscillator amplitude, 92, 142, 156
$\mathbf{E}_R$	Raman signal amplitude, 142, 156
$\mathbf{E}_S, E_S$	signal amplitude, 13, 61, 68
E_{sy}	y-polarized signal amplitude, 78
E_t	transmitted probe amplitude, 14, 22, 164
E_2	amplitude of wave number 2, 68
$\mathbf{E}(\mathbf{r},t), E(t)$	electric field, 2, 32, 43, 60
$\varepsilon, \varepsilon'$	sign of energy difference, 81
$\varepsilon(t)$	amplitude of nonlinearly radiated wave, 51
ε_σ	fractional RMS fluctuation, 157
η	initial phase of $\boldsymbol{\beta}$ vector, 37
$|f_c\rangle$	continuum wave function of state f, 188
$|f_d\rangle$	discrete wave function of state f, 188
f	oscillator strength, 190, 231
F	photon flux, 20, 172–177
F_Q	generalized force on coordinate Q, 18
$|f(\mathbf{r})\rangle$	wave function for state f, 173
$f(x)$	coherence correction to Lamb dip, 72
$|g\rangle$	wave function for state g (ground state), 117, 161
G	velocity distribution function, 48
g_α	degeneracy of level α, 81
$g_D(v)$	distribution function of population difference, 67
$g(v_+), G'(\Delta)$	velocity detuning distribution function, 216
G_L	radial wave function, 173
$\langle g|\bar{u}_\alpha|n\rangle$	dipole moment matrix element for g-n transition in polarization α, 116, 186

Appendix: Symbol Glossary–Index

$g(\Omega,\Omega_0,\Omega_D)$, $G(\omega-\Omega_{ab})$	distribution function of resonant frequencies, 48, 198
$g_{D0y}(v)$	distribution function for 0-y population difference, 77
$\tilde{G}(t)$	unitary β transformation matrix, 38
Γ	rate of elastic collisions, 205
$\Gamma(t)$	instantaneous ionization rate, 43
Γ_B	three-level saturation linewidth, 81
Γ_{bb}	decay rate of ρ_{bb}, 82
Γ_{eff}	width of Lamb dip, 95
$\Gamma_{tg}^{(M)}$	M-photon transition rate between state g and t, 177
Γ_N	three-level coherence linewidth, 81
$\Gamma_{fg}^{(N)}$, $\Gamma^{(N)}$	N-photon ionization rate of state g, 172, 177
$\Gamma_{ft}^{(N-M)}$	$N-M$ photon ionization rate of state t, 117
γ_ϕ	pure dephasing rate for two-level system, 33
Γ_t	transverse relaxation rate, 177
Γ_{tg}	transition rate between states t and g, 161, 166
Γ_{uv}	decay rate of ρ_{uv}, 58
$\Gamma^{(N)}$	N-photon transition rate, 21
Γ_l	longitudinal relaxation rate, 177
Γ_2	transverse relaxation rate $1/T_2$, 24
$\Gamma_{\rightrightarrows}$	linewidth for copropagating laser beams in three-level systems, 82
$\Gamma_{\leftrightarrows}$	linewidth for counterpropagating beams, 82
$\mathcal{H}$	Hamiltonian operator, 31
$\mathcal{H}_I(t), \mathcal{H}_{I_{ba}}$	interaction Hamiltonian, 32, 50, 51, 53,
$<t|\mathcal{H}_I|g>$	matrix element of $\mathcal{H}_I$ between states t and g, 161
$\mathcal{H}_{\omega_1-\omega_2}$	interaction Hamiltonian resonant at $\omega_1-\omega_2$, 117
$\tilde{\mathcal{H}}_R$	relaxation Hamiltonian, 32

Appendix: Symbol Glossary–Index

$\mathcal{H}_0$	isolated atom Hamiltonian, 32
I	intensity, 2, 142
I_+, I_-	intensity of + or − wave, 68
I_B	incoherent background intensity, 156
$I_h(t)$	heterodyne intensity, 62, 156
I_{LO}	local oscillator intensity, 156
$I(\omega_2)$	intensity at frequency ω_2, 62
$I(\omega_1-\omega_2)$	CARS intensity as a function of $(\omega_1-\omega_2)$, 135
$I_R, I_R(t)$	Raman-radiated intensity, 142, 156
I_{sat}	saturation intensity, 69
I_s	signal intensity, 62
$J_0(x), J_1(x)$	Bessel function of order 0 or 1, 199
$\mathbf{k}$	wave vector of photoelectron, 172
$\mathbf{k}$	wave vector of light, 164
$\mathbf{k}_i, \mathbf{k}_0, \mathbf{k}_1, \mathbf{k}_2$	wave vector of indicated light wave, 12, 31, 60, 164
k_p	wave vector of radiating polarization, 60, 61, 131
K_1, K_2	correction factors, 60
$\kappa, \kappa_0, \kappa_1, \kappa_2$	absorption constant of indicated wave (unsaturated), 2, 68, 69, 71, 81, 136
$\kappa', \kappa'(\omega,\omega), \kappa'(\omega_1,\omega_2)$	saturated absorption parameter, 14, 68
κ_s	absorption constant for signal wave, 136
ℓ	sample length, 19
ℓ_c	coherence length, 131
ℓ_{opt}	optimum sample length, 136
λ	detector response parameter, 156
λ	wavelength, 7
λ_c	collection efficiency, 158
$|m\rangle$	state m wave function (intermediate state), 162
M	number of photons to reach level, 96
$\mathcal{M}, m$	molecular mass, 48, 96
$\bar{\mu}(t), \mu, \bar{\mu}, \mu_{ab} = \langle a|\bar{\mu}|b\rangle$	transition dipole moment operator, 2, 30, 52, 53

Appendix: Symbol Glossary–Index

$\tilde{\mu}_\alpha(t)$	α spatial component of $\tilde{\mu}$, 117
$\boldsymbol{\mu}_{un}$	transition dipole moment between states u and n, 166, 167
$(\tilde{\mu}_{un}\tilde{\mu}_{nv}\tilde{\mu}_{vm}\tilde{\mu}_{mu})$	product of four-dipole operators: spatial components specified separately, 59
$\mathcal{N}$	number density of ensemble, 3, 45, 59, 68, 135
N	order of multiphoton interaction, 22, 172, 177
n_e, n_0	extraordinary and ordinary index of refraction of uniaxial crystal, 228
$N_{\text{eff}}, N_{\text{exp}}$	experimentally determined order of multiphoton interaction, 175, 176, 179
n_{mix}	index of refraction of mixture, 230
$n(\omega_3,\theta)$	index of refraction at frequency ω_3 and angle θ, 227
N_x, N_y	number of density of atom x or y, 231
n_2	nonlinear index of refraction, 150
$\Omega, \Omega_{ab}, \Omega_{ac}, \Omega_{tg}$	resonant frequency for indicated levels, 34, 74, 116, 164, 166, 186, 216, 223
Ω_0	center resonance of split states, 70
Ω'	three-level resonance, 81
$\Omega_\pm$	Autler–Townes resonance frequencies, 70
Ω_b	energy of level b in frequency units, 236
Ω_D	Doppler width, 48, 67, 165
$\hat{\Omega}_{ij}$	complex frequency parameter for ij transition, 59
Ω_Q	frequency of Raman mode Q, 117
Ω_R, Ω_{rg}	Raman frequency, 115, 233, 239
Ω_{xi}, Ω_{yi}	frequency of ith transition of species x or y, 231
$\Omega_1, \Omega_2, \Omega_3$	correction factor frequency arguments, 60
ω	laser frequency, 2, 66, 164
ω_s	signal frequency, 132
ω_x	crossover resonance frequency, 74, 75
$\tilde{P}$	projection operator, 51

Appendix: Symbol Glossary–Index

$\mathcal{P}^{(2)}$	second harmonic power, 227
$\mathbf{P}, \mathbf{P}(t)$	dielectric polarization density, 2, 45, 48, 67, 203, 204, 208, 210
$\mathbf{P}^{(3)}$	third-order polarization, 49, 60, 198
$\mathcal{P}_H$	heterodyne signal power, 149
$P_i^{(2)}(\omega_3)$	ith spatial component of second-order polarization at frequency ω_3, 227
P_i	ith spatial component of P, 4
$\mathcal{P}_L$	laser power, 158
P^{NL}	nonlinear polarization, 156
p_Ψ	probability of state Ψ, 31
$\mathbf{P}_Q, P^Q$	coherent Raman polarization of mode Q, 18, 156
$P_\alpha^Q(\omega_s)$	α spatial component of ω_s Fourier component of $\mathbf{P}^Q(t)$, 118
$P_\alpha^Q(t)$	α spatial component of $\mathbf{P}^Q(t)$, 117
$P_x^{(3)}$	x-component of $\mathbf{P}^{(3)}$, 54
$\mathcal{P}_1, \mathcal{P}_2$	power of laser beam 1 and 2, 149
$\langle \delta \mathcal{P}_k^2 \rangle$	mean square fluctuation in kth laser power, 157
$\langle \delta \rho_\sigma^2 \rangle$	fluctuation in σth portion of response, 157
ϕ	phase shift between pulses, 213
ϕ	angle of $\mathbf{R}$ in rotating reference frame, 94
$\phi(v_z, \tau_1, \tau_2, \tau_3)$	Borde velocity phase factor, 58
$\Phi_\Delta(t)$	angle of $\mathbf{R}$ vectors at time t of subensemble with detuning Δ, 201
$\tilde{Q}$	projection operator, 51
$\mathbf{q}$	wave vector of Raman mode, 154
q	Fano parameter, 188
Q	Raman coordinate, 18, 117
$|r\rangle$	Raman resonant state wavefunction, 117
$\mathbf{R}'$	Bloch–Feynman vector in fixed reference frame, 36
$\mathbf{R}$	Bloch–Feynman vector (in rotating frame), 38, 94, 211
$\mathbf{R}_\Delta$	$\mathbf{R}$ for subensemble with detuning Δ, 201, 214, 215

r_e	classical radius of electron, 231
$\mathbf{R}_I$, $\mathbf{R}_{II}$	**R** vector after interaction I and II, 211
$\mathbf{R}_\perp$	component of **R** perpendicular to $\hat{3}$ axis, 201
ρ	response of photodetector, 156
ρ	density matrix, 30–63
ρ, $\rho(E)$	density of states, 2, 165
$\rho_{ab}^{(1)}(\omega_1)$	off diagonal element of density matrix first order in E and oscillating at ω_1, 44
$\rho_{ab}^{(3)}(\omega_1)$,	ω_1 component of ρ_{ab} third order in E, 44
ρ_B	response due to background light, 157
ρ_{vv}^e	equilibrium density matrix element for state v, 58
ρ_H	heterodyne response, 157
ρ_{LO}	response due to local oscillator, 157
ρ_Q	depolarization ratio of Raman mode, 125, 154
ρ_R	response due to intensity radiated by Raman mode, 15
$\rho_{ab}^{(3)}(2\omega_1-\omega_2)$	component of $\rho_{ab}^{(3)}$ at frequency $2\omega_1-\omega_2$, 44
$\rho_{uv}^{(n)}$	term in ρ_{uv} proportional to $\mathbf{E}^n$, 56
ρ_1	response due to single photon detection, 157
S	signal, 200
S/N	signal-to-noise ratio, 157
sF	optical Stark shift, 177
$\hat{\sigma}_{(N-M)}$	$N-M$ quantum ionization cross section, 177
$|\Psi\rangle$	general wave function, 29
σ	electronic hyperpolarizability, 150
σ_N	N quantum ionization cross section, 20, 177
σ_Q, $\sigma(\omega_1-\omega_2)$, $d\sigma_{\alpha\beta}/d\Omega$, $d^2\sigma_{11}/d\Omega d(\omega_1-\omega_2)$	Raman scattering cross section, 3, 20, 130
$|t\rangle$	wave function of multiquantum resonant state t, 161, 174, 176, 199

Appendix: Symbol Glossary–Index

T	$T_1 = T_2 = T$, 199
T_a, T_b	lifetime of state a or b, 33, 43, 69, 79, 210
t_d	time between interactions, 201, 211
T_{eff}	effective temperature, 182
$<t|\mathcal{H}_1|g>$	matrix element of $\mathcal{H}_1$ between states t and g, 161
t_p	pulse length, 196
t_t	transit time, 95
T_w	decay time of population difference, 72
T_1	longitudinal relaxation time, 33
T_2^*	inverse of inhomogeneous linewidth, 197
T_2	transverse relaxation time, 20, 33, 43, 166
t_2, t_3	time of second or third pulse, 201, 206
τ_a, τ_0	alignment and orientation decay times, 79
τ_1, τ_2, τ_3	interaction times in Borde formalism, 56
$\theta, \theta', \phi, \phi'$	beam interaction angles, 131, 203, 228
Θ	pulse area, 38, 201, 204, 211
Tr	trace of quantum mechanical operator, 30, 45
u, u(t)	component of **R** vector in rotating frame, 37, 39, 40
u(∞)	value of u(t) in steady state, 40
$U_0(x)$	unit step function, 81
u'	component of **R'** in nonrotating frame, 36
v	atomic velocity, 164
v, v(t)	component of **R** in rotating frame, 37, 39, 40
v_x, v_y, v_z	components of atomic velocity, 48
v_0	RMS thermal velocity, 48
v_+, v_-	velocity of resonant atoms, 216
v(∞)	value of v component of **R** vector in steady state, 40
v_{ab}, v_{ac}	velocities of atoms in resonance with indicated transition, 74

$<v\|\bar{\chi}^{(3)}_{\alpha\beta\gamma\delta}(-\omega_p,\omega_0,\omega_1,-\omega_2)\|u>$	nonlinear susceptibility tensor matrix element, 58
V_E	configuration interaction, 188
$\hat{V}_\varepsilon$	total matter-field Hamiltonian in Friedman–Wilson–Gordon formalism, 51
v'	component of **R′** in nonrotating frame, 36
w, w(t)	component of **R** corresponding to population difference, 37, 39, 40, 215
$w(\infty)$	value of w in steady state, 41, 67
$w(0)$, $w^{(2)}(0)$, $w^{(2)}(\omega_1-\omega_2)$	Taylor–Fourier components of w, 44
w_Δ	value of w for subensemble with detuning Δ, 207, 209,
δw_Δ, Δw	population difference variation, 95, 216
$\delta w'_{II}(\Delta)$	fluctuating part of w after interaction II as function of Δ, 212
w_0	minimum radius of laser beam, 63
w'	component of **R′** corresponding to population difference in nonrotating frame, 36
Z	partition function, 149
ζ	noise equivalent power of detector, 157
$\hat{1}, \hat{2}, \hat{3}$	basis vectors for rotating reference frame of Feynman–Vernon–Hellwarth vector model, 36

INDEX

A

Absorption, 3, 49, 67
  multiquantum, 4, 21, 161
    Doppler-free, 23
  saturated, 4, 13, 62–103
  two-photon, 151
Ackerhalt–Eberly condition, 182
Acoustooptic modulator, 219
Adiabatic following, 39, 40
Alignment, 78
Antiholes, 104
Anti-Stokes, 115, 236
Applications, 110, 148, 189
Atom
  dressed, 39
  Rydberg, 170, 192
Attenuation, 2
Autler–Townes effect, 39, 70
Autoionizing states, 187, 237

B

Background free geometry, 78
Balmer line, 16, 100
Beer's law, 2
Benzene, 150
Birefringence, 78, 92, 139, 227
Birefringent media, 132
Bloch–Feynman vector (**R**), 36, 37, 45, 195, 199
Borde diagram, 55–57
Born–Oppenheimer approximation, 116, 125
Bottleneck, 104
Boxcars, 20, 132, 155
Broadening, 167
  homogeneous, 14, 47
  inhomogeneous, 14, 15, 47, 66, 81, 154, 163, 199
  pressure, 171
  transit time, 70, 96, 211
Build-up cavity, 89

C

Calcium, 184
CARS, *see* Coherent Anti-Stokes Raman Spectroscopy
Cesium, 178
CH_4, 97
Channels, 174
Chi-three ($\chi^{(3)}$), 121
Coherence, 31, 71, 80, 102
Coherence length, 189
Coherent Anti-Stokes Raman Spectroscopy (CARS), 19, 52, 62, 116, 118, 119, 131, 132, 232
  cw, 138
  minimum, 138
  multiplex, 137
Coherent Raman Spectroscopy (CRS), 13, 17, 115–160

Index

Coherent Stokes Raman Spectroscopy (CSRS), 19, 116, 118, 132, 233
Coherent superposition state, 29
Collisions, 205
Color center lasers, 11
Combustion diagnostics, 148
Computer memory, 110
Condensed phases, 48, 103
Contour integration, 49
Convective derivatives, 48
Conventional Old-fashioned Ordinary Raman Spectroscopy (COORS), 116, 158
COORS, see Conventional Old-fashioned Ordinary Raman Spectroscopy
Correction factors, 60
Crossover resonances, 73, 198
Cross section, 149
Crystal classes, 122
CSRS, see Coherent Stokes Raman Spectroscopy

D

Degeneracy, 81
Degenerate four-wave mixing, 111
Density matrix, 29, 48
Dephasing, 163
Detected intensity, 62
Detected signal, 60
Detuning, 67
Diagnostics
  combustion, 148
Diagrammatic techniques, 54
Dichroism, 16, 78, 92, 139
Dielectric polarization density, 1, 4, 45
Dielectric susceptibility, 1
Dipole moments, 199
  operator, 53
Dispersion, 67, 131, 227
Doppler effect, 13
  transverse, 96
Doppler width, 66, 81, 154, 165
  residual, 85
Doppler-free multiquantum absorption, 23
Doppler-free resonances, 58
Doppler-free three-photon interaction, 169
Doppler-free two-photon absorption, 164
Dressed atom, 39
Dye, 10
Dye laser, 4

E

Echo
  photon, 24, 201
  stimulated, 24, 206–211
  stored, 205
Effective operators, 50, 166
Effective two-level interaction Hamiltonian, 51
Einstein convention, 47
Emission, 3
Ensemble, 31
Excimer lasers, 13
Expectation value, 30
Experimental methods, 84, 217
Experimental results, 97
Extended two-level model, 177
Extracavity techniques, 85

F

Fano–Beutler line-shape, 188
FID, see Free Induction Decay
Field ionization, 191
Fluence, 180
Fluorescence, 106
  intermodulated, 89
  saturated, 84, 89
Fluorescent samples, 152
Four-level system, 58
Four-wave mixing (4 WM), 4, 19, 45, 119, 147, 209
  degenerate, 111
  two-photon resonant, 54
Free Induction Decay (FID), 24, 195, 196, 211, 217
  first order, 196
Frequency shifting, 219

G

Golden rule, 22, 161

H

Hamiltonian, 31, 32
  effective two-level interaction, 5
  total matter-field interaction, 51

Index

Hänsch–Borde saturation spectrometer, 86
Heterodyne detection, 217
Heterodyne signal, 143
High-resolution spectroscopy, 151
Higher-Order Anti-Stokes Scattering (HORAS), 21
Higher-Order Stokes Effect Scattering (HORSES), 21
Higher-order sum generation, 230
Hole, 14, 47, 66, 73, 103
Hole burning, 80, 104
Homogeneous broadening, 14, 47
Homogeneous linewidth, 15, 24, 104, 212
HORAS, see Higher-Order Anti-Stokes Scattering
HORSES, see Higher-Order Stokes Effect Scattering
Hydrogen, 98, 170

I

Incoherence, 31
Index of refraction, 81
Infrared spectrophotography, 25, 237
Inhomogeneous broadening, 14, 15, 47, 66, 81, 154, 163, 199
Instantaneous ionization rate, 43
Intensity, 2, 62
Interaction
  Doppler-free three-photon, 169
Interferometric techniques, 90
Intermodulated fluorescence, 89
Iodine, 98, 154
Ionization
  field, 191
  multiphoton, stepwise, 183
  multiquantum, 172
  resonant, 176
IR detectors, 192
Isolation, 86, 87
Isotope shift, 100

K

k-space spectroscopy, 155
Kleinman symmetry, 125
Kramers–Kronig relationships, 3, 68

L

Lamb dip, 66, 68, 74, 85
  inverted, 84
Lamb shift, 100
Lasers
  color center, 11
  dye, 4
  excimer, 13
  lead salt diode, 12
  optically pumped three-level, 83
  pulsed, 221
  pump, 14, 68
  tunable, 4
Laser-induced fluorescence line narrowing, 106
Lead salt diode lasers, 12
Light scattering, 3
Line shapes, 96, 143, 166
Linewidth, 7, 104, 213
  homogeneous, 15, 24, 104, 212
Liouville equation, 30
Literature, 26
Littrow condition, 7
Local field, 2
Local oscillator wave, 62, 92, 156

M

Magnetic permeability, 1
Magnetization, 1
Maker–Terhune notation, 46
Master equation, 34, 58
Maxwell–Boltzmann velocity distribution, 48
Mixing
  four-wave, 19, 147, 186
    degenerate, 111
    two-photon resonant, 54
  nonlinear, 185, 230
  three-wave, 119
  wave, 232
Molecular dissociation, 179
Momentum space, 154
Moving grating effects, 45, 104, 109
Multiple resonance effects, 54, 57, 179
Multiquantum dipole moment operator, 52
Multiquantum dissociation, 181
Multiquantum ionization, 172
  resonant, 176, 179

Multiquantum saturation spectroscopy, 101
Multiquantum transitions, 50

N

Nonlinear mixing, 185, 230
Nonlinear polarization density, 45
Nonlinear resonances, 60
Nonlinear sources, 25, 226
Nonlinear susceptibility, 4, 45, 69, 77, 81, 118, 149, 163, 232
Nonlinear susceptibility tensor elements, 46, 122
Nonresonant ionization, 172
Nonresonant term, 118
Nuclear response, 125
Nutation, 217

O

OARS, see Optoacoustic Raman Spectroscopy
OHD–RIKES, see Optical Heterodyne Detected RIKES
Operator, 30
 dipole moments, 52
 effective, 50, 166
 multiquantum dipole moment, 52
 projection, 51
Optical bistability, 110
Optical coherent transients, 24, 195–225
Optical (or IR) double resonance, 102
Optical Heterodyne Detected RIKES (OHD–RIKES), 142
Optical heterodyne detection, 62
Optical nutation, 199
Optical pumping, 78
Optically pumped three-level laser, 83
Optoacoustic Raman Spectroscopy (OARS), 21
Optogalvanic effect, 89
Order N, 172
 experimental (N_{exp}), 176
Orientation, 78
Oscillation
 Raman, 232
 stimulated, 119
Oscillator strengths, 189
Output frequency resonance, 187

P

Parametric oscillator, 12
Parity, 22
PARS, see Photoacoustic Raman Spectroscopy
Permutations, 47
Phase, 31
Phase matching, 12, 61, 228
Phase modulator, 219
Phase switching, 221
Phase velocity, 2
Photoacoustic Raman Spectroscopy (PARS), 21
Photoionization, 42
Photon correlation, 174
Photon echo, 24, 201
Point-group, 121
Polariton, 154
Polarizability, 3
Polarization spectroscopy, 16, 73, 76, 90, 100, 104, 109
Polarization switching, 221
Porphyrin, 105
Position dependence, 45
Poynting vector, 2
Pr^{3+}: LaF_3, 205
Pressure broadening, 171
Probe beam, 14, 76
Projection operator, 51
Pseudofield vector, 36, 199
Pulse area, 38
Pulsed lasers, 221
Pump laser, 14, 68

Q

Quantum beats, 205
Quantum defect, 171
Quantum electronics, 26
Quasi-continuum, 162

R

Rabi flopping frequency, 38
Rabi frequency, 34, 195
Raman Induced Kerr Effect Spectroscopy (RIKES), 20, 139–159
Raman mode, 116
Raman oscillation, 232

Index

Raman resonances, 187
Raman scattering, 115
Raman scattering cross section, 3, 20, 130
Raman shifting, 232
Raman susceptibility tensor, 4, 53, 126
Ramsey fringes, 25, 92, 211
Rayleigh resonance, 103, 109
Recoil effect, 96, 98
Reference frame, 53
Relative cross sections, 174
Relaxation, 24, 32, 39, 155, 199
  transverse, 24
Relaxation time, longitudinal, 33
Residue theorem, 49
Resolution, 138
Resonances, 37, 47
  crossover, 73, 198
  Doppler-free, 58
  nonlinear, 60
  Raman, 187
Retroreflectors, 88
Rf saturation techniques, 102
RIKES, see Raman Induced Kerr Effect Spectroscopy
Rotating reference frame, 36
Rotating-wave approximation, 34, 45
Rotational Raman resonance, 235
Rubidium, 170
Rydberg atom, 170, 192
Rydberg constant, 100, 184

S

Saturated absorption, 4, 13, 62–112
Saturated dispersion, 69, 85
Saturated fluorescence, 84, 89
Saturation intensity, 69, 74
Saturation spectroscopy, 66, 214
Schroedinger equation, 30
Second harmonic generation, 226
Selection rules, 166
Self-focusing, 150
Sensitivity, 156
Servo system, 228
Shot noise, 157
Shutters, 221
Signal amplitude, 13, 68, 78
Signal-to-noise ratio, 156
Slowly varying complex signal amplitude, 61

Sodium, 184
Spatial dispersion, 132
Spectral diffusion, 107
Spectroscopy
  high-resolution, 151
  k-space, 155
  multiquantum saturation, 101
  polarization, 16, 73, 76, 90, 100, 104, 109
  saturation, 66, 214
Spherical tensor representation, 125
Spontaneous cross section, 130
Spontaneous emission, 33
Spontaneous Raman scattering, 17, 143, 154, 158
SRG, see Stimulated Raman Gain
SRL, see Stimulated Raman Loss
SRS, see Stimulated Raman Scattering
Stark effect, 218
  optical, 23, 52, 167, 176
Steady state, 40
Stepwise multiphoton ionization, 183
Stepwise transitions, 82, 161
Stimulated echo, 24, 206–211
Stimulated emission, 3
Stimulated Raman effect, 25
Stimulated Raman Gain (SRG), 4, 20, 54, 116, 119, 147
Stimulated Raman Loss (SRL), 20, 145
Stimulated Raman oscillation, 119
Stimulated Raman Scattering (SRS), 20, 62, 145
Stokes scattering, 115
Stored echo, 205
Strong field limit, 70
Strontium, 198
Successive approximation, method of, 43
Sum frequency generation, 4, 226, 230
Superposition, 31
Symmetry, 121, 189
Synthetic aperture technique, 93

T

Taylor–Fourier coefficients, 43
Techniques
  diagrammatic, 54
  extracavity, 85
  interferometric, 90
  rf saturation, 102

Tensor
  Raman susceptibility, 4, 53, 126
  two-photon, 186
The Inverse Raman Effect (TIRE), 20, 116, 119, 145
The Raman Induced Kerr Effect (TRIKE), 20, 116, 119
Theory, 29–64
Third-harmonic generation, 231
Third-order nonlinear susceptibility $\chi^{(3)}$, 46, 58, 121, 196
Three-wave mixing, 119
Time dilation, 217
TIRE, see The Inverse Raman Effect
Total matter-field interaction Hamiltonian, 51
Transient, 155
Transit time broadening, 70, 96, 211
Transitions, 80
Transverse decay rate, 199
Transverse relaxation, 24, 32, 43
TRIKE, see The Raman Induced Kerr Effect
Tunable laser, 4
Two-level system, 29
Two-photon absorption, 151
Two-photon optical precession, 223
Two-photon resonant four-wave mixing, 54
Two-photon resonant third-harmonic generation, 54
Two-photon tensor, 186
Two-photon transient, 223
2π pulse, 38

V

Vacuum ultraviolet, 25
Vector
  Bloch–Feynman (**R**), 36, 37, 45, 195, 199
  Poynting, 2
  pseudofield, 36, 199
  wave, 155
Vector model, 34
Velocity changing collisions, 96
Virtual intermediate states, 161
Volume effect, 175

W

Wave-front conjugation, 110
Wave mixing, 232
Wave vector, 155
Wave-vector matching condition, 14, 61, 130, 227

X

XUV Anti-Stokes, 26, 236

Y

Yajima's Rayleigh resonance formula, 49

Z

Zeeman effect, 104, 167
Zero phonon lines, 104